VINTAGE GUNS

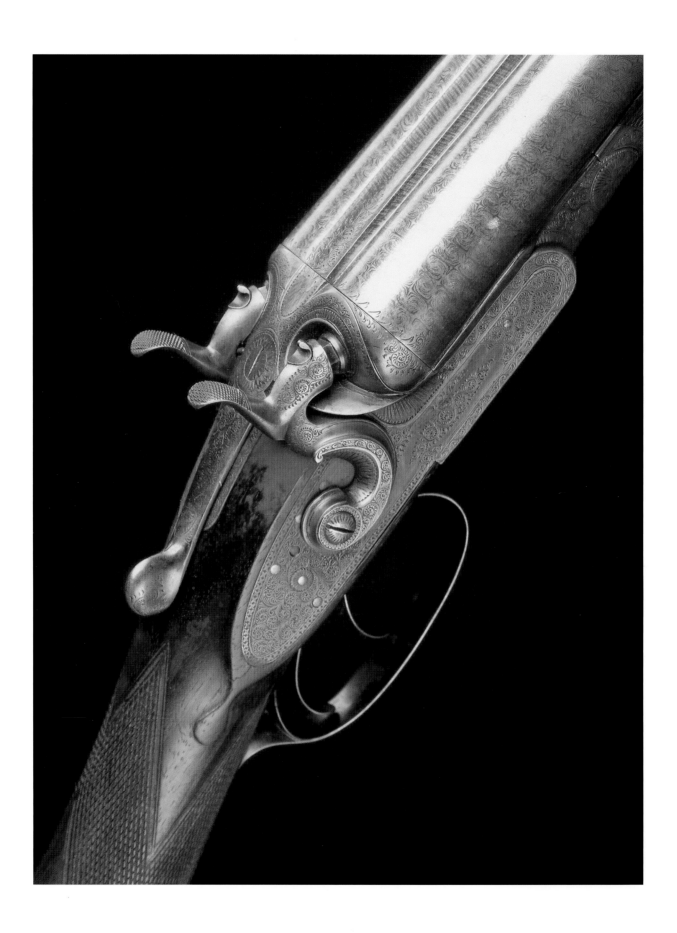

VINTAGE GUNS

Collecting, Restoring and Shooting Classic Firearms

Diggory Hadoke

Skyhorse Publishing

Published in the USA by Skyhorse Publications Inc, 2008

www.skyhorsepublishing.com

10 9 8 7 6 5 4 3 2 1

Library of Congress Cataloging-in-Publication Data

Hadoke, Diggory.
Vintage Guns: Collecting, Restoring & Shooting Classic Firearms / Diggory Hadoke.
p. cm.
Includes bibliographical references.

ISBN-13: 978-1-60239-198-7 (alk. paper)
ISBN-10: 1-60239-198-X
1. Firearms–Catalogs. 2. Firearms–Collectors and collecting. I. Title.

TS534.7.H34 2008
683.40075–dc22
2007015909

Printed in Singapore

CONTENTS

PART III: USING VINTAGE GUNS

PHOTOGRAPH ACKNOWLEDGMENTS

The author wishes to thank a distinguished list of contributors for generously allowing the use of their original photographs to illustrate this book. They are:

Michael Yardley: 16, 30, 33, 42, 43, 50, 75, 77, 82, 84, 97, 124, 146, 148, 157, 161, 163; Andrew Orr of Holt and Company, (who deserves a medal): 2, 10, 33, 34, 40, 42, 46, 64, 67, 68, 72, 74, 75, 76, 81, 82, 84, 88, 90, 92, 95, 109, 112, 120, 127, 128, 138, 190; Gavin Gardiner (of Gavin Gardiner Ltd and Sotheby's): 64, 101, 168; Steve Sidki (of Gunstock Blanks): 70; John Foster (of John Foster Gunmakers) 50; Richard Purdey (of James Purdey & Sons): 43, 164, 166, 170, 171, 173, 175, 176, 182; Russell Wilkin (of Holland & Holland): 119; Barry Lee Hands: 44, 174; William Powell (Gunmakers): 37; Jason Kane: 184; Bonham's: 63, 95; Damasteel Ltd, Sweden: 117; Keith Chard: 131.

Unless otherwise credited, all original photographs were taken by the author. Further images were scanned from old documents, advertising literature or out-of-print publications.

'Shoot to enjoy yourself and in such a way that you are a pleasure to shoot next to all day, and you will find you can shoot all your life and when you can no longer shoot, you will always be missed and your memory remain golden.'

T.D.S & J.A Purdey

DEDICATION

In memory of E.G 'Joe' and Lilian Mister

With the help of readers' comments, I intend to continue to learn...

A FEW WORDS OF THANKS

As I put finger to keyboard to write this in anticipation of my first book appearing on the shelves in Britain, the United States and possibly elsewhere, I find myself reflecting on how it all came about. Of course, those who write are not always the experts on their chosen topic; much of what we write we have absorbed from others and I confess to feeling a little fraudulent in this respect.

People see a name in a magazine or on the cover of a book and in their mind's eye form the image of an expert. In the case of this book I could name a dozen people who have more intimate knowledge and a greater depth of understanding on much of the subject-matter than me. However, not all experts feel the inclination to put their thoughts on paper and, by not doing so, much of their knowledge would remain unavailable to the outside world or be lost to history. Where I have made factual errors, I welcome correction. If I have offended, I ask forgiveness – for no malice was intended.

My thanks for help with the writing of this book must go to Mike Yardley for his corrections; his keen eye for detail and factual accuracy have helped reduce the error count and focus the scope of the content. Mike's writing provided me with the impetus to attempt a book myself, and his encouragement, genuine expertise and generous access to his wonderful library have been invaluable.

Going back in time, I owe a debt of gratitude to Joe Anson, Joan and Mike Wise, late of Ludlow School, who found some promise in a terribly unco-operative pupil, while others despaired. Through a combination of patient encouragement and a willingness to fight fire with fire, these three ensured that, whatever talent for English and History may have been resting behind a façade of idleness, had no choice but to get some exercise.

Dave Mitchell candidly recounted the tales of his time as a Purdey gunsmith and Richard Purdey generously took time to answer my correspondence concerning modern Purdey training and other matters. Robin Nathan, late of Purdey, so obliging and enthusiastic when allowing me access to the 'day-books' and the Long Room at South Audley Street, is also due credit. Philip and Will Beasley helped with the history of the pigeon magnet and provided an excellent excuse to spend yet another day in the pigeon hide 'researching'.

Nick Holt, Chris Beaumont, Ralph Paschen, Paul Attfield-Downes, Andrew Orr and Howie Dixon of Holt's deserve my thanks for their patience and good humour while allowing me free access to their stock for photographing and for their permission to use Andrew's own excellent photographs for illustrative purposes. I am also indebted to Matthew Smith and Ian Andrews of Christie's, and to the staff at Holland & Holland. Gavin Gardiner, Mick Shepherd, James Marchington, Dan Cote, John Hargreaves, Alan Myers, and John Farrugia, Barry Lee Hands, Patrick Hawes and Paul Roberts have all played their parts.

Finally, my thanks to Federica Comini for tolerating my unfathomable hobby and to my mother, Jan, for tolerating a lot more, for much longer. I must also gratefully acknowledge those kind friends and sportsmen who helped by reading manuscripts, adding suggestions or correcting errors, which as a result no longer lurk in the text to embarrass me. These include Pete Woodgate, Robin Knowles, habitués of the *Double Gun* BBS, Peter Jones, Don Amos and Merlin Unwin.

Lists of thanks are tedious so I'll end mine now. I would be grateful for any critical feedback from readers. Writing this has been a learning process and I hope that readers' comments will ensure I continue to learn as the years go by.

Diggory Hadoke, London

www.vintageguns.co.uk

April 2007

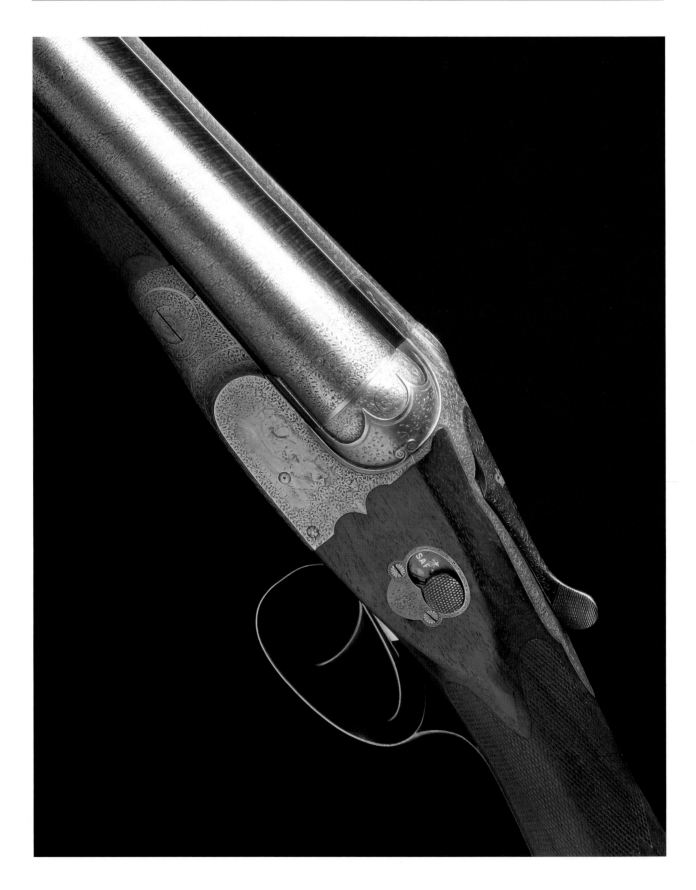

FOREWORD
by Michael Yardley

I first met Dig Hadoke when he came to see me to have a gun fitted – his much-used and much-loved 1889 Purdey. It was soon apparent that he was an enthusiast not only for English guns, but for the old style of shooting sport. He had his share of driven days; probably more than he could really afford but he also had a passion for rough shooting and pigeon shooting and mucking about in a field with an old air-rifle or .22. It transpired that we had much in common too; absent fathers, somewhat wayward childhoods spent wandering the fields with a gun, and later years wandering the world trying to escape the conforming tyranny of modern life. I have a few years on Dig, but we soon became firm friends and frequent shooting companions.

As is evident in this volume, Dig has a thirst for knowledge and, a great gift: an easy, fluid, writing style, combined with clear thinking and an eye for detail. When I read the first draft of this book, I was impressed. In fact, parts of it bowled me over. The personal anecdotes gave the book character, but the technical information concerning the development of English guns gave it real authority. This was not the usual glib or second-hand regurgitation. It was fresh, interesting and useful. Dig gave an overview based on experiment and hands-on comparison. He understood what he was writing about. He put things into context, and created time lines. He was especially good at comparing the opinions of the experts of the past; men such as Hawker, Payne-Gallwey and Major Burrard, as well as adding his own, always practical, advice.

The book is a treasure trove for any 'practical eccentric' interested in shooting and/or collecting old guns. It is written by a real shooting man who finishes his own gunstocks (beautifully) and who I have seen take half a dozen different guns on a driven day 'because it will be interesting to compare them'. Within its pages you will discover how a traditional gun is made (including some most interesting interviews with gunmakers), how it might be repaired or proofed. You will discover tips on buying at auction, and a consideration of what offers value and what does not (it is unlikely that you will ever see these subjects better analyzed). I have contributed a few short notes on shooting and gun fit. There are excellent sections by the author comparing new guns to old, hammer guns to hammerless, and boxlocks to sidelocks.

This is a book that deserves a place on every shooting man's bookshelf. It will certainly be bought by those who are enthusiastic about older guns – it's a 'must have' for them – but it deserves the widest possible audience. It is a book written by a professional teacher and, consequently, it is not surprising that it informs so well. If you want to understand sporting shotguns and their development, this book will help you greatly. The consideration of characters, people who make and use guns, is especially entertaining. The material on Richard Arnold (recently described in *Shooting Times* as 'mad as a bat'), Phil Beasley the pigeon shot and inventor of the 'magnet', and Percy Stanbury (arguably the most elegant shot of all time) caught my attention.

All things considered, Dig has a rare talent for writing about guns and shooting. He is not just considering the object, but he has added the human dimension. This is his first book, but I am quite sure that it will not be his last.

Witham, Essex.

Spring 2007

PROLOGUE

There are many books on guns by gun experts, many books on shooting by shooting instructors and many books on the countryside by eminent naturalists. In fact, every subject seems to have a guru able to tell us what we need to know. This is not such a book. I undertook to write it not as an authority but as an enthusiast, in the hope that some of this enthusiasm is catching and some of what I have learned may be of use to others of like mind.

The book aims to open up the mysteries of the world of old guns, too often secretive and impenetrable, full of pitfalls for the unwary. It offers practical advice on gathering the necessary knowledge and contacts to buy and sympathetically renovate guns for use in the field. It outlines a practical strategy for using such guns in modern shooting contexts and draws on my own experiences and the anecdotes of my sporting companions.

The book is designed for easy reference. Technical information is provided in an accessible format rather than requiring the reader to hold a degree in engineering to understand it. It is meant for the shooter, rather than the scholar or historian.

This is an unashamedly personal book about the pursuit of shooting and gun collecting as a passion, one kindled in deepest childhood or perhaps even before; who knows why we love what we do?

My advocacy of shooting old guns in the modern world casts a deliberate backwards glance in the face of modern shooting practices and harks back to simpler days, simpler pleasures and the quality of experience gained through the acknowledgement of tradition in our sport.

My passion is for old guns and their continued use in the field and my argument encompasses the 'quality of the impractical', as well as making real, practical arguments in building a case for their superiority to the ubiquitous machine-made over/under of the modern shooting field.

The book covers all aspects of collecting shotguns to use, on the budget of the mere mortal rather than the Captain of Industry. No place here for gold-encrusted exhibition pieces, just appreciation of bygone quality surviving in the modern world.

My views are often controversial and at odds with the conventional wisdom currently governing shooting matters. However, all the points made are, I hope, backed with a clear rationale. Historical, evaluative and qualitative issues mingle to give an essence of what using old guns can bring to your shooting.

In summary, this book is aimed at those who shoot, who may shoot or have shot. Those who shoot old guns will find much common ground; those who do not should find much of interest and those who have always wondered about this eccentric but compelling world will find a practical way into it.

At the back of many a gun cabinet lurks an inherited but neglected old shotgun. If this book encourages a few owners to bring some of them back into use, then I will be happy it was not a purely academic undertaking.

This is an account of my own pursuit of my abiding passion: the lessons I have learned, the knowledge I have gained and the beliefs I have developed. It contains practical and factual information as well as opinion and seeks to be not only a useful tool for those seeking information but a compelling read for those whose interest is more general.

The author with childhood friend Jason Kane (right). We started our adventures with guns as 8-year-olds.

The author in the famous Long Room at Audley House with a Purdey 4-barrel 20-bore.

INTRODUCTION

My mother thought guns would make me violent. I wasn't allowed them. So, I became a four-year-old thief and stole my friends' toy guns and hid them under my bed.

By age eight I had a .22 BSA Super Meteor that I had to kneel to shoot, because I could not stand and hold it for very long. I don't remember ever wanting anything from my parents from that point on, except pellets. Lots of them. I hung tin cans in trees and shot them and I walked long and hard with my little gun in search of Shropshire rabbits. I delighted in the thrill of shooting rats from my perch on the top of the chicken shed as dusk fell. I shot sparrows out of the hawthorn bushes and once, in the days when England still had elm trees, I even shot a rook circling 40 yards above a rookery (right through the chin!).

I shared my passion with my earliest shooting friend Jason Kane. His father, John, a big, enthusiastic exiled Irishman, would take us on 'expeditions' after rabbits and treated our boyish obsession with manly seriousness, proffering wise words on safety and responsibility, yet maintaining our enthusiasm. In short, we respected him and wanted to emulate his sportsmanship.

At twelve I asked my father for a shotgun (a nice little .410 double like I saw in the *Shooting Times* classifieds, I thought). 'There's one in the cupboard,' wrote my father, from Germany where he had found a new life with a new woman.

I looked and found a Webley & Scott 700 series 12-bore with 30" barrels and so my shotgun shooting life began. I shot my first 'right and left' at wild duck with this gun a couple of years later – by torchlight as they came off a flooded field – while out rabbiting with a school friend!

From childhood I have sustained an interest in old guns. My influences were largely what psychologists call 'nurture', I suppose. My father was a country GP, ex-army officer and gun collector, though his interest seemed to extend only as far as taking the guns down occasionally to 'sniff', as my mother put it. His invitation to shoot each Boxing Day with Lord Boyne never overly worried the local pheasant population. He told me once that he was better with a machine gun!

However, his guns were there for me to find and handle when he was out (no security required in the late 1970s). As luck would have it, my primary school headmaster, Tony Sheppard, was an avid gun collector and gifted artist who was able to carve new wooden grips for his revolver handles out of mahogany and often brought recently acquired firearms into school to show us. I remember a double elephant gun particularly vividly.

How times have changed: my cousin, Rufus, was recently chastised by a hysterical teacher just for having the *BASC Guide to Shooting* in his possession at school. In modern Britain guns are bad – 'Haven't you heard of Columbine?' was the, admittedly topical, admonishment to his mother. Fortunately, Gaby has the sense and balance to play such blinkered stupidity with a straight bat.

I grew up in rural Shropshire, one of four, we made our own fun and were out in the fields from morning to nightfall, running in for food at set times and then off again to explore, fight, destroy, construct, wander and ponder in the world about us.

So, what should one make of 'nature'? – for 'nurture' cannot explain all. My brother Paddy developed no interest in guns at all. Nor did Toby, nor Susie. This little obsession was mine alone.

It must have always been in me somewhere and, though it has faded at times, it has always lingered and returned; this interest in the hunt. The evocative smell of gun-oil, metal and old wood, of powder and dogs and leather. These have held me for as long as I can remember and they hold me still.

The author aged thirteen with a Webley & Scott 12-bore and four of the unluckiest hares in Shropshire.

A box of Curtis & Harvey's extra coarse gunpowder for duck and coast shooting.

PART I

SHOOTING & COLLECTING

Awaiting the arrival of wild ducks in early autumn with a Purdey 12-bore which first killed game in the 1889-1890 season.

Why do we shoot?

Human beings are odd creatures; their behavior is unfathomable at times. Some things we do because we have to (eat, sleep, work). Other things we do to please other people (paint the hall, iron shirts, shave) but we frequently maintain an area of life that generates a momentum of its own. It becomes its own *raison d'être* and for everybody it is a little different, with a different spark that set it off, now long forgotten.

Some people stand on platforms and note train numbers, some make model ships out of matches, yet others swap partners and video the proceedings for their own amusement. I, and I suspect most of you reading this, shoot. What we do and what separates us from 'the others', binding us together in a loose coalition, is the inexplicable, addictive draw of the chase, the fascination with the implements of it and the execution of it in good style.

Whether a 'high-roller' shooting 50 days of driven pheasant each year or a working father of three who can only manage the odd pot-hunt after rabbits, whether a competitive clay shooter or a solitary wild-fowler, our differences do not detract from this link we have to one another.

Just go to a Game Fair to see the variety of human 'breeds' on show. The tattooed Yorkshire pigeon shooter, the trilby-wearing syndicate member with a country law practice, the teenage airgun-toting rabbit slayer, the tweedy classes with their own estates and the ferreters from estates of an entirely different nature. All in the same place, all linked together, all different – yet all the same. This is who we are.

Shooting is a great leveller. In this picture of a between-drives coffee break can be seen a fireman, a game-keeper, a successful businessman and the parent of a Hollywood movie star. It is impossible to tell them apart – for all are equal in the shooting field.

'Have you ever wondered why you shoot?' This is a question often asked by those who do not understand the compulsion of the sport. We all endeavour to justify our sport and ourselves. How many of these have you tried:

- I love being outside in the fresh air.
- I have to shoot vermin; it is a duty and a necessity.
- Shooting my own wild meat is healthier – it is all 'organic' and I know where it comes from.
- Wild animals are happier than farmed animals, they have a happy life and, as top predator, I harvest them naturally. Shooting is more ethical.
- It is good exercise.
- It gives me an excuse to walk the dog.
- It gets me out of the house.
- I like the challenge.
- It is sociable.
- It is elemental, part of my nature as a genetically developed hunter.
- Shooting helps the rural economy.
- It fills the freezer.

I'm sure you could add innumerable more rational reasons for picking up a gun and going outside. But by doing so, don't we miss the point?

We could do any number of things that fulfil the above personal benefits just as well and we could as happily shirk the 'responsibility and duty' side of the argument as we do all the other duties and responsibilities of society that we leave to other people.

But we shooters don't really understand why we shoot, other than that something, some ghost perhaps, makes us want to. It is a part of who we are and we are a part of it. We are all haunted houses. Some are haunted by the ghosts of Walsingham, whose shooting bankrupted him, or Ripon, whose shooting eclipsed his political career and was the defining force in his life, for he is remembered for little else.

Others may be visited by those whose pursuit of excellence and style drove the development of the shotgun, created the shooting styles that form the cornerstone of its use in the field or recorded the history of sporting gun development: James Purdey, Col. Peter Hawker, Sir Ralph Payne-Gallwey, Joe Manton, Robert Churchill, Percy Stanbury or 'Charles Lancaster' (H.A.A. Thorn), or any number of anonymous sportsmen who may have contributed what they were able on a smaller, more localized scale.

The ghosts of gunmakers past like Manton, Purdey, Lancaster, Beesley and Greener; whose invention and dedication drove sporting gun development to the improbable perfection it attained, are my own personal ghosts. Perhaps yours are adventurers with the gun such as Teddy Roosevelt, the hunting-obsessed US President, or Colonial tiger-slayers like Jim Corbett (who trained my grandfather in jungle warfare). Perhaps you are visited by a dead grandfather, who would come in smelling of damp tweed, spaniel and gunpowder and sit by the fire to tell you of his day in the field.

Whoever it was who provided you with the spirit that inhabits your daydreams and fuels your pursuit of shooting, they are passing on the elemental drive that inhabits certain men, to seek out the gun, the rifle, the quarry and the outdoors. We are what others have been and what future generations may be. We will, one day, be those same ghosts.

Returning to shooting – a modern phenomenon

In times past, shooting people progressed along more predictable lines than is generally the case today. Some are still fortunate enough in their sporting lives to take the traditional route and develop from children with air guns to youths with proper air rifles, to teenagers with small-bore shotguns. Then comes the first 12-bore on the 17th birthday and gradually the lone shooter moves into more social shooting circles as age and experience progress.

This traditional, sporting timeline served many purposes. The child learned safety and field craft. He learned about the quarry. He learned about the guns, the dogs and the etiquette of shooting, progressively. Only after a long apprenticeship in the ways and lore of the gun and the shooting field would he find himself a regular Shot in a line of Guns facing driven birds. In time, he might even assume the role of shoot captain and pass on what he learned to others. And so, the story would continue.

It is true that for many Victorians and Edwardians, their early twenties and thirties involved postings to far-flung parts of the Empire, but the sporting opportunities for many in such situations were enhanced rather than reduced. India and Africa offered unrivalled sport of every conceivable kind: waterfowl, antelope or tiger simply

Geoffrey Boothroyd (above left) and H.A.A. Thorn (Charles Lancaster): writers separated by a generation who have both left an enduring legacy.

replaced the mallard, rabbit and red deer as quarry.

The early 21st century is a very different world. Boys still grow up on farms but rather than stay and tend crops or beasts, they go to university and become accountants, TV executives, advertising or marketing managers, or any of a number of other urban-based professions. Those with roots in the countryside and shooting sports lose their

Lord Walsingham was not only famous as a game shot, he was also a passionate naturalist and authority on minute moths known as microlepidoptera. His passion for shooting eventually cost him his fortune, including the site of the Ritz Hotel and two country estates totalling over 10,000 acres.

interest as they spread their wings and go in search of the outside world. But they come back.

Like the middle-aged 'born-again biker', the prodigal shooter reaches a point in life where his mind returns to the passions of his youth. This can be an unwitting process; it came home to me that I was in this transitional phase one day when I was discussing the idea of replacing my flashy silver BMW coupé with some kind of grubby Jeep.

'My boot is always clogged up with decoys, cartridges, wellies, gun-cases and stuff', I heard myself complain to my mother on the telephone. 'You are becoming the boy you once were', she replied. She was right.

I was a little taken aback. My priorities had shifted. No longer was my car a symbol of my cool or my taste in engineering, to be polished and driven to the limit, it had become a means of getting me to a muddy field with all my kit for a day's shooting.

This fact was reinforced later that month, when I found myself driving said BMW across a stubble field, sills scraping and skirts squealing, sports suspension protesting all the way and a local farmer keeping a conspicuously safe distance from the obvious madman driving to the middle of his stubble field in a sports car. All of this in order to set up my 'pigeon-magnet' and hide.

At sixteen, I used to drive an equally unsuitable vehicle – a Honda SS50 moped with cased Webley and Scott strapped to the pillion seat to go rabbiting on local farms. Back to the future indeed!

Statistically, I am in good company. There are many of us out there with an urge to return to our sporting roots. Gone is the competitive drive of youth – the need to be the strongest, the coolest, the toughest, the most popular. Gone too is the need to conform (though I was never much good at that). The time is ripe for reflection on the value of shooting as a hobby or sport and the interest and value to be had from our re-emergence as shooting people.

Let us not buy the newest offering from the production line of some modern sporting gun manufacturer and kit ourselves out in head-to-toe technical shooting suits like some new-Russian oligarch on a corporate jolly. Let us take it slowly and re-absorb the pleasures of the old gunroom and other simple pleasures: sporting birds, good company and ancient shotguns.

Let us allow our eccentricity to have free rein and bring it to life in the modern shooting context. You may seem out of place at first but you are not alone. As you practise your own peculiar brand of shooting and collecting, you will start to recognise like-minded folk as they come into sharper focus. As surely as you seek their company, they will find yours.

Continuing an English tradition: Guns make their way into the mist for the first drive on a November morning.

The BMW 325 Sport Coupe: not the ultimate shooting vehicle.

A proud history of Practical Eccentricity

The 'practical eccentric,' as I shall call him, is generally not to be found as a member of the gun trade 'proper' (though 'practical but eccentric' does serve as a fairly accurate description of many of the minds involved in the development of the shotgun. It requires a certain gifted obsessiveness to come up with so many ways of making a tube spit lead). He is probably somebody whose interest in shooting began in childhood and his natural inquisitiveness led him to want to find out more.

Some, if not most, people are satisfied to just 'do' and are not prone to analyse, research or explore. They will buy the latest gun, as they would purchase a car or a fridge, then use it, clean it, put it away and follow the herd to the clay shoot or game shoot. They enjoy their sport and good luck to them but they are more 'normal' than the practical eccentric. They are modern consumers and participants in a sport, much like golfers.

I like reading. I am especially fond of reading old books in poor condition by odd people. Over the years I have picked up numerous such titles from Game Fairs and second-hand bookstores. I am also fortunate that by happy coincidence my friend Mike Yardley has probably the best library of shooting-related material in the world, including the entire collection of the late Chris Cradock, whose own thoughts are often scribbled in the margins of many of the books. Britain has a great tradition of producing writers on subjects related to guns and shooting, America no less so.

These men (for they are all men) indulged in their passion, hobby, or however you wish to see such things with a single-minded dedication to their purpose. Some were rich and had the time and money to indulge their fancy. Others, ordinary in every outward sense, were scholarly, inventive and inquisitive. Whether designing the ultimate punt gun, like Sir Ralph Payne-Gallwey, producing home-made camouflage coats or building their own punts like Richard Arnold or designing an 'infernal contraption' to lure wily pigeon into shot, like Phil Beasley – all have made a contribution to the establishment and maintenance of the tradition of the practical eccentric in British sporting history.

Lt-Col Peter Hawker

MacDonald Hastings called him 'The father of game gun shooting'. Hawker was a soldier who served with distinction in the Peninsular War and a diarist, author and theoretician in all matters shooting. His life spanned the development of sporting guns from flintlock to breechloader and he was a close friend of Joseph Manton, the great gunmaker of his time. James Purdey (who worked for Manton) famously said that gunmakers of his generation would all have been 'a parcel of blacksmiths' but for Manton's influence. I have lost count of the times Hawker notes in his diary that he had taken delivery of a new gun from Manton. He certainly had a good collection. Among guns still in existence that belonged to Hawker are those in the (recently disbursed) late William Keith Neil collection. Hawker's diaries form an enduring record of a remarkable life at an important

Hawker was the son of an Irish mother and an elderly father of military background, who had both wealth and royal connections. He owned lands in Middlesex and had an estate in the loveliest part of the Test Valley in Hampshire, which provided fabulous game shooting and chalk-stream trout fishing. Robin Knowles, whose help was invaluable in my research of Hawker, reflected: 'My old grandfather used to say: *Youth and money should go together, but they so rarely do.*' Well, for Hawker they coincided most fortunately and this 'eternal boy' was able to indulge his passions to his heart's desire.

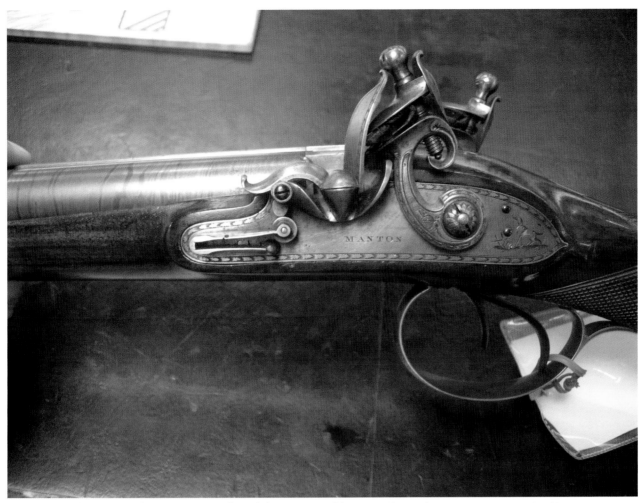

This Manton flintlock is of Hawker's era, though he converted his own Manton's to percussion cap ignition.

Left: This Smith percussion muzzle loader is the type of weapon that rapidly replaced the flintlock during Hawker's lifetime.

time in the progress of game shooting and firearms development.

Hawker was reputedly a fabulous shot and claims (one of many claims of uncommonly good consecutive scores) to have shot 13 jack snipe, when aged just sixteen, without a miss – using a flintlock! He kept detailed records of his bags and exploits and championed the merits of the finest in gunmaking for the whole of his life. His legacy is a very eccentric English attitude to the pursuit of game, some conventions of which he established. Other, now abandoned, practices of Hawker's seem most odd to modern sportsmen, such as chasing partridges to exhaustion on horseback and shooting birds on the ground as happily as those in the air. His quarry included starlings, landrails, herons, woodpeckers and pretty much anything else that ran or flew (including his own dog on one occasion while shooting in France) – and he was not averse to organized poaching when the mood took him.

One would expect a wounded officer, with a hip bone shattered by a musket ball in battle, to rest at home in fear of his life. Instead, Hawker got his servants to push him around in his bath chair so he *continued to drive out almost everyday, taking my gun, and killing (from the carriage) redwings, fieldfares, blackbirds, larks &c. Today, among other things, I killed several snipes'*. You may not share his enthusiasm for the bagging of songbirds but you have to admire his verve considering there were no antibiotics and such wounds as his were prone to become fatal. Hawker's wound took over a year to heal but he was out shooting almost everyday anyway.

His views and exploits are saved in his volumes *Instructions to Young Sportsmen in all that Relates to Guns and Shooting*, which were printed in ten editions from 1814 to 1854, and in his diaries, first published in 1893. MacDonald Hastings in *The Shotgun* (1981) tells us that Hawker composed the epitaph etched on the tombstone of his gunmaker and friend, Joseph Manton. The wording gives us an idea of the sensibilities of the man and the value he placed on Manton's gunmaking skills.

'In memory of Mr Joseph Manton, who died universally regretted, on the 29ᵗʰ June 1835 age 69. This humble tablet is placed here by his afflicted family, merely to mark where are deposited his mortal remains. But an everlasting monument to his unrivalled genius is already established in every quarter of the globe, by his celebrity as the greatest artist in firearms that the world produced,

as founder and father of the modern gun trade, and as a most scientific inventor in these departments, not only for the benefit of his friends, and the sporting world, but for the good of his King and country.'

Hawker's diaries, sympathetically edited (some would say misleadingly) for publication by Sir Ralph Payne-Gallwey, in 1893, provide a fascinating insight into the early development of what were to become modern sporting guns and the countryman's pursuit of game, long before the days of the great Victorian and Edwardian driven shoots. One cannot help but wonder about the veracity of Hawker's claims as a Shot, (I am suspicious that he tended to discount shots he later decided were not 'decent chances' when calculating his consecutive kills) but later writers are generally in agreement that, while Hawker's 'extreme range' may have actually been around 35 yards (Hawker says not) he did record his exploits largely accurately.

Teasdale-Buckle, writing in *Experts on Guns and Shooting* in 1900 asserted 'we must accept Col Hawker's extraordinary long shots as well within killing range, otherwise he could not have made the great number of consecutive kills that he constantly – and we believe – truly records'.

An interesting margin note in the book from the respected shooting coach Chris Cradock records his thoughts in pencil:

'Too right, I don't believe it either – about 1⅓ shots per head killed (in his lifetime) – too good!'

Other commentators suggest that Payne-Gallwey deliberately attempted to paint a picture of Hawker as a kind of solitary game shooting superman. Actually, he seems to have been a gregarious and sociable shooter: an engraving of him shooting at Longparish shows him in the company of Joe Manton and others in pursuit of game in 1827. Payne-Gallwey stands accused of having deleted the names of Hawker's companions and ascribed their portions of the bag to him alone. Eric Parker produced a version of Hawker's diaries in 1931, based on the unexpurgated script from which Payne-Gallwey worked, which he claims to be a more accurate representation.

To understand Hawker's exploits, we must understand his shooting environment. Robin Knowles describes it thus:

'Much of his shooting was that typical of his day; repetitious 'bum shooting', mostly at grey partridges and hares, either from horseback or on foot over rock-

Robin Knowles, a noted firearms historian, an authority on Peter Hawker – and a practical eccentric of the first order.

Below: The 1852 Lancaster base-fire breech loader was way ahead of its time, patented one year before Hawker's death.

steady pointers and setters. These were often worked in pairs; and in relays on big days. This form of shooting is unlike modern driven shooting or shooting birds put out of coverts by dogs such as spaniels. Shooting in this old-fashioned/traditional manner is somewhat akin to down-the-line shooting. You have to be able to read your dogs and likely game-holding areas on the terrain in conjunction with the prevailing atmospheric conditions. Then, when game is located and held by the dogs, to approach with gun held at the ready and mind emptied of everything but a single-minded concentration on the impending shot. Then, when the game rises, following it with one's eyes, bringing the gun up through the bird from below, at the same time avoiding shooting over it, and firing at the killing distance of 30-35 yards, taking care to avoid a long shot at the wounding distance of 45 yards or so. Shooting in this manner and using black powder you soon get used to the 'woomph' sound and a certain amount of white smoke, and you've nearly always shot your quarry dead. The technique was known to me in Ireland as 'hopping the beggars'. Not at all difficult once you get the hang of it.'

Robin recalls having learned the technique on the bogs of County Mayo, many years ago, with a single 12-bore muzzle-loader by McKnight of Dublin, so he speaks from experience. He also believes Hawker's guns well-suited to the job, having no choke but, typically of 14-bore, with barrels friction and relief bored, loaded with 1¼oz of soft No.7 shot with best felt wadding and 2¾ drams of progressive-burning black-powder of far better quality than that now available. This would have provided good, dense patterns and equalled the performance of a modern choke-bored gun with 30g loads, for range. From the 1820s onwards, Hawker shot with Lancaster and Westley Richards guns, generally doubles of 14-bore. He quickly abandoned flint in favour of percussion ignition and had his favorite Mantons 'Old Joe' (a double 20-bore of 1807) and 'Big Joe' (his single 4-bore duck gun) converted to percussion cap.

The story of Hawker is inextricably linked with that of his gunmaker, Joe Manton. Manton is rightly famed for his development of the flintlock in the first thirty years of the 19th century. However, many of Manton's contemporaries produced guns of equal quality in Hawker's time. Durs Egg, William Moore, Samuel Nock, William Parker and William Smith are just a few of the 'artists in firearms' whose memory is not perhaps as golden as Manton's but whose work lives on as testament to their skills. Manton's

friendships with Hawker and Lord Mountjoy enabled him to evade the normal social conventions and actually accompany them shooting; a contravention of the rigid social barriers that at that time prevailed. Manton made good use of such friendships and patronage and applied his entrepreneurial skills and magnetic personality to building a famous business.

His guns' qualities, apart from the sheer quality of their material construction, lie in the superb balance and handling, combined with reliably fast ignition. Many of his guns still in existence are worn out after years of consistent use in the field. Many were converted to percussion ignition, expensively and properly, with new percussion breechings rather that with the cheap and commonplace drum and nipple screwed into the touch-hole.

Whatever the truth about the two men, Hawker and Manton contributed to one another's reputations and provide modern readers and shooters with a lasting romance from the dawn of modern game shooting and with the guns developed for the purpose.

Richard Arnold

Richard Arnold was very much the 'practical eccentric' in all that he did and wrote. His views and practices were outside the conventions of his time and he was at once forward-looking and nostalgic. He embraced the qualities of the repeating shotgun and defended it in the face of attack by traditionalists, whose prejudice led to all sorts of spurious criticism of this type of gun when it first appeared at British game shoots. Yet Arnold was as partial to using a muzzle-loader for his pigeon shooting as he was a semi-auto. His instruction in the making of pigeon decoys is as bizarre as it is fascinating and shows the modern shooter, with money and plastic gizmos readily available, how the inventive mind of the austere post-war era applied itself to the challenges of our sport.

Arnold was a Mancunian of Scottish ancestry, domiciled in London and he worked for a time for Keely Wilson, a long defunct importer of gun and shooting accessories. In 1952 he was one of the founding members of the Muzzle-Loaders Association of Great Britain, whose excellently eccentric motto was 'Claret and Black Powder'.

So much for eccentricity but what of practicality? The 'Practical Eccentric' is not a madman, totally out of step with the world around him. He has good reason for doing what he does and he can back his unorthodox choices and practices with studied and compelling argument.

Richard Arnold was technically knowledgeable about the things that made up his world. He understood ballistics, he learned natural history and he studied field craft. He was familiar with the mechanical workings of old guns and of new ones and he knew the advantages of both. In all his writing he makes perfect sense, yet what he advocated, at the time, was a long way out of step with his contemporaries. He wrote (in 1950) of his captivation with muzzle loaders:

> 'to the gun lover – who likes to handle and appraise good weapons, to feel their balance, marvel at their locks, and understand the craftsmanship that has been embodied in the precision steel-work, and their polished stocks – there is not a gun made today, even by the best gunmakers, which can rival a muzzle loader built by one of the old craftsmen.'

He understood the quality of the piece and how to use it to best advantage.

My favorite picture of Arnold is on the front of his book *The Shooter's Handbook*. It shows him in an old felt hat, a kind of tatty Macintosh, tightly knotted necktie and holding a muzzle loading hammer gun. He looks superficially like a cross between Little Bo Peep and the village idiot. Yet his body language is all focus and poise as he addresses a target. His style was as personal, un-selfconscious and practically eccentric as the rest of him and his appearance belies his sharp intellect and deep

Richard Arnold embraced both the percussion muzzle-loader and the semi-automatic.

knowledge of subject. Colin Willock, writing in *Shooting Times* in 2005 affectionately referred to Arnold (who he knew) as 'as mad as a bat'.

Arnold's most interesting book, *The Shooter's Handbook,* resulted from discussions in a pub in Red Lion Square, Holborn, among these early members of the M-LAGB, who believed it would be a good thing to make available a *vade mecum* of essential and practical shooting information. It is a wonderful window into the post-war re-discovery of elemental shooting implements at a time when much of what was known could have passed out of living memory.

In a modern world where style rules over substance, this humble but exacting man of strong mind stands out from the many as the best one can read on guns and field-craft. He writes about what he actually did; making decoys, camouflaging his old shooting coat and refurbishing old guns, loading cartridges and so forth. His work is original and captivating and is as well worth reading today as it was when published, perhaps more so for the snapshot it gives of the austere shooting world of the 1950s, a much less chronicled shooting era than the exploits of the Edwardian upper classes.

Frederick Beesley

The history of brilliant innovators and inventors in the world of gunmaking is long and distinguished and it is a risky business to pull one name to the fore. In those days before TV, what did these men do when the working day was over? It seems that a great many of them sat down and worked out new ways of making guns more efficient. Their names are carried on in their inventions, as we shall see later. The Purdey bolt, the Scott spindle, the Greener cross-bolt and the Anson & Deeley boxlock carry the names of their inventors for posterity. John Robertson was arguably the inventor of the most important refinements of the breechloader with his single trigger, his ejector and his superlative over-and-under patent of 1909, though he was ably assisted by Bob Henderson in devising the latter, so shouldn't take all the credit.

So why Frederick Beesley? He is much less a household name than many of his contemporaries. I recently competed in a sporting clays event for vintage guns and saw a nice sidelock of around 1885 and asked the owner 'Is that a Beesley you have there?' He was quite affronted and proudly told me it was actually a Purdey, which he clearly felt was a step up in class. Purdey may be the name on the lock plates but it was Beesley's brain that conceived this most famous of actions.

Beesley was known as 'The Inventor to the London Trade' during his lifetime and is responsible for some groundbreaking designs that we still see in use today. Burrard noted that 'Mr F. Beesley was one of the most fertile inventors the gun trade has ever known' in *The Modern Shotgun* (1931). Like John Robertson, Thomas Perkes and others like them, Beesley invented and developed a great deal in a relatively short period of time and the gun trade made wide use of his inventive genius.

Beesley was reputedly a native of Oxfordshire and, at the age of fifteen, left his farming background to apprentice

Frederick Beesley (1846-1928),
inventor of the Purdey Sidelock

with Moore & Grey in Old Bond Street, London. Moore & Grey were top rank gunmakers, with Manton connections, and young Beesley clearly had an aptitude for the work. After completing his apprenticeship and working at various locations within the London gun trade he joined Purdey in 1869 as a stocker and remained at the firm until 1878, when he started his own firm off Edgware Road.

While at Purdey's, Beesley must have worked on the ideas that were to ripen and eventually reach fruition as the 'Purdey Sidelock' that we know today. In December 1879, shortly after setting up in business, he approached his old employer James Purdey with his design and sold it to

him in January 1880, for a fee of $97. Beesley accepted an offer of royalty payments of five shillings on the first 200 guns made. Purdey started to make the guns and quickly acted on the option to buy out the rights for a further single payment of $170 and closed the deal in November 1880.

Beesley probably found the Purdey money very useful in establishing his new business and he eventually prospered, inventing all the time. We know of the relationship between Purdey and Beesley because it is well-recorded but it is likely that he invented other mechanisms that were patented by others that we do not know of. What we do know is that before his death in 1928 he patented designs for single triggers, ejectors, safety devices for boxlocks, self-cocking actions and an ingenious over-and-under action called 'The Shotover'.

Perhaps his finest legacy is the 'Purdey' sidelock, which has been in continuous production in all calibres and bores, as a rifle and shotgun respectively, since the year of its invention, and for his ejector work on developing the Perkes-originated 'over-centre' principle, which forms the backbone of what is today known generically as the 'Southgate Ejector'.

It is remarkable that 19th century gunmakers with little or no formal education managed to become so adept at grasping the principles of engineering and physics required to devise the efficient mechanisms for operating firearms that they did. Of all of them, Frederick Beesley, the farm boy from Oxfordshire, stands out as a unique talent and a testament to the ingenuity latent in many human beings, which can emerge when the mind is applied obsessively to a single subject.

Sir Ralph Payne-Gallwey

Sir Ralph Payne-Gallwey carried on in the tradition of Colonel Peter Hawker in the pursuit of practical eccentricity; indeed he was the logical choice to edit Hawker's diaries, which he did. There is a well-known photo of Sir Ralph in his gunroom surrounded by bows, shotguns, some enormous punt guns (of which he was a famous exponent) and other shooting equipment. This was his element and though he was not a gunmaker by trade, but a 'gentleman', he upheld that honourable English tradition of the amateur expert.

Coming from the perspective of the end-user rather than the manufacturer and retailer has its advantages.

Payne-Gallwey applied his mind to equipment and tactics that reflected his needs. He shot all the time and he shot everything from bows to punt guns. This gave him plenty of time to develop ideas on what he wanted his equipment to be like and to test his ideas and designs rigorously in the field. To this day we have no better design for a cartridge bag than the pigskin Payne-Gallwey model and if you have a little brush for scrubbing the chambers of your shotgun, it is actually called a 'Payne-Gallwey brush' and was another of his inventions. He is also on record as trying to sell Purdey his idea of a stock-mounted game counter, unsuccessfully by all accounts.

Payne-Gallwey was an avid wildfowler and another of his inventions was the double-barrelled punt gun based on the patent screw breech of Henry Holland. This gun was built for Payne-Gallwey by Holland & Holland and tested at their Kensal Rise shooting ground in March 1885. The gun was remarkable in that it was designed for the fall of the first hammer to set in motion the fall of the second, slightly delaying the second discharge. This way the second charge would hit birds taking off after the first shot hit them on the water. Each shot discharged 20oz of lead. He is reported to have once bagged 70 widgeon with one double discharge. Sadly, this gun no longer exists; it was sold to the founder of WAGBI (now BASC) Stanley Duncan and later converted to become two single guns. Duncan wrote of this conversion (or desecration) in *The Field* in 1921.

Payne-Gallwey, who died without a male heir

Sir Ralph Payne-Gallwey in his gunroom at Thirkleby Hall. His gravestone reads: 'He nothing common did, or mean'

(his son was killed in Flanders in 1916), was also a prolific author, contributing the 'Shooting' component of the Badminton Library with Lord Walsingham, as well as penning half a dozen books of his own: *High Pheasants in Theory and Practice* and *Letters to Young Shooters* being the best known. He was also a great pundit on shooting matters and punt-gunning in the 1880s and 1890s, with articles appearing regularly in *The Field* and *Land and Water*. One of his many eccentric exploits was the building of Roman siege engines (on which he was a published authority), which he test-fired over the Menai Straits.

Philip Beasley

Sir Ralph Payne-Gallwey invented monstrous punt guns for the taking of wildfowl in the 19th century and Philip Beasley spent the closing years of the 20th century devising the ultimate device for the taking of the woodpigeon. Philip, like Frederick Beesley, comes from Oxfordshire farming stock and though the two men do not share the same spelling of their surname, they certainly share a flair for invention.

Game shooting has generally been seen as an organized and expensive sport but the shooting of woodpigeon was for many years considered purely as vermin control and undertaken by shooters acting with the blessing of the relevant Government Ministry. The Government even provided cartridges to approved personnel in the post-war years. 'Pest officers may now refund such registered shots half the cost of cartridges which they purchase from their own gunsmiths' (wrote Richard Arnold in 1956).

Woodpigeon are truly wild birds and cannot be driven towards a line of guns with any regularity. Bringing them into shot has always been an art, and as such is seen by many game shooters as beyond their capabilities.

Nowadays, pigeon shooting is a commercial sporting enterprise and the casual shooter can buy a day in which he is guided to success and may shoot large numbers of birds. He can also set himself up in a field and draw pigeons into shot more successfully today than he has ever been able to do in the past. The single most important factor in this state of affairs is the 'Pigeon Magnet'.

Philip Beasley is a lifelong shooting man (named in *The Field* in 2003 as one of the top 10 game shots in Britain) who found there was commercial potential in taking people pigeon shooting, something he had been doing for years in an un-paid capacity. In the mid-1990s, the standard method of drawing pigeon in to shot was with decoys. Some contraptions existed for providing movement, such as manual flappers but many clients, who were unused to pigeon shooting, found them a difficult quarry and new methods of attracting birds to the decoy pattern were always being sought.

The initial inspiration came from some Italian clients who demonstrated a device they used at home to attract song birds. Idle speculation about whether it might work on pigeons followed and Phil cobbled together a manually operated rotary attractor based on the Italian model but big enough on which to mount two dead pigeons. The major component was a bicycle wheel and, though rudimentary, it worked well enough to warrant further reflection.

Philip had been an engineer in his formative years and had worked on racing cars for the March Racing Team, repairing F1 and F2 cars for drivers like Nikki Lauder and Ronnie Peterson. Making crashed cars race-worthy under the pressure imposed at race meets requires the ability to improvise solutions to engineering problems and it proved a good training ground. He later graduated to working on prototypes for new components and this expertise is what he applied to developing a fully mechanical rotary pigeon decoy.

The first attractor was constructed and prepared for trial without a great deal of expectation. A friend was asked to try it out and duly took it off for a day in the pigeon hide. It worked immediately and an excited call came through

Phil Beasley continues the British tradition of eccentric innovation in the shooting field and has devised some bizarre yet highly effective devices for the pigeon shooter.

to Philip saying the birds were coming in from half a mile away and mobbing the machine even though the shooter was not fully concealed and that he was shooting a good number of birds as they approached. It heralded the start of a revolution.

Initially the 'Pigeon Magnet', as Philip trademarked his machine, aroused scepticism in the shooting press and the claims made about it were met with derision. At first only Philip's shooting clients saw the full benefits but this helped shore up the successful shooting-guide business.

The machine is brilliantly simple. A multi-pronged spike secures it in the ground and from this a single shaft stands upright. Upon this is mounted a compact 12v motor and a swivel with mounting points for two 'arms'. These flexible arms extend outwards and may be adjusted for angle. At their extremity is an adjustable three-pronged cradle. A dead bird is skewered on the middle prong and the wings spread and secured on the outer prongs. When the machine is activated, the whole thing moves with a rotary motion and the birds appear to be circling to land.

Philip had some 'Pigeon Magnets' made up by a

The 'Pigeon Magnet' – a simple device that has revolutionised pigeon shooting.

local engineering firm and advertised them for sale in shooting magazines. However, the big break came with the launch of a promotional video. This showed the Magnet being set up and the shooting that followed and it was shown at the premier annual event in Britain for shooting sportsmen: the CLA Game Fair. The public was hooked, the shooting press crowded around to write articles and business took off. Phillip sold 4,000 Pigeon Magnets in the year that followed. The next year trade increased to a turnover of $990,000.

Philip has continued to improve the original device and it has had many incarnations, some of which were over-complicated and, consequently, were dropped. Others superseded the original and the 2005 version is lighter, easier to set up and transport and more flexible. Philip also continues to invent: various mechanical flappers, gliders and peckers have been trialed and marketed since 1995 and the stream of invention shows no sign of drying up.

The quest continues as Philip works to stay one step ahead of the wily wood pigeon. 'Pigeons are natural survivors', he says, 'they grow wary of recognized dangers and the shooter has to continue to work at varying the movement in a pattern of decoys in order to attract them in to shot.' He now trades in Britain as *UK Shoot Warehouse* but for many, Philip Beasley will remain best known as the inventor of the 'Pigeon Magnet', and rightly so.

Competitive and non-competitive shooters

I have recently started shooting in sporting clay competitions in Essex. The events are largely well-organized and well-attended by men in skeet vests. It is a world for which the game shot and the pigeon hide inhabitant is largely unprepared.

'What's that?' inquired a female shooting acquaintance of mine recently, addressing a large and successful competitor. 'Thirty-four inch Perrazzi', says the bearer gruffly, holding it proudly in both hands. The message could not be clearer. It could not be further away from the mindset of the practical eccentric.

For here we have the gun as a masculinity extension. It goes with wide wheels, of the 'Carlos Fandango' type featured in old Hamlet cigar adverts, with bleached, trophy girlfriends and 4-wheel drive pickups. Another world.

This is not the natural environment of the practical eccentric. Don't get me wrong, the Essex clay shoot is not a bad thing: in fact, I like it. I meet my friends there, I get practice with my guns in the 'dead time' between seasons and I know I am doing an injustice to my fellow competitors with my unkind comments. However, although we shoot together we are not the same. A friend put it thus: 'Some shoot to feed their ego, others because it is a quest for self-knowledge'.

Competition in the shooting world is not new. In the Edwardian era it was commonplace at lunchtime, to read out the number of birds each Gun had shot. It was not unheard of for a gentleman to pick up another's birds if he had not shot well: the cause of more than a few competitive contretemps on the grouse moor! It was also considered important to be able to shoot quickly as well as accurately. For this, Lord Walsingham was especially famed.

There are definitely two different types of shooter: those who shoot to win competitions and for whom shooting expresses a competitive drive – a need to win – and those

Percy Stanbury: The last man to win a major open championship with a side-by-side. Stanbury is pictured here with his Webley & Scott 500 12-bore, with full-choked 30" barrels, inlaid with commemorative oval plates for each competition won with it. He was still using it with Alphamax No.4s at the age of 84 for shooting high pheasants at East Meon in Hampshire. Stanbury also used a W&C Scott Monte Carlo Special live pigeon gun, again, choked full in both barrels.

who shoot because it pleases them and for no other reason. Though I am admitting my own weakness here, we in the latter camp may shoot competitively but we will never be driven by competition. True competitors are poor losers, occasional ones phlegmatic in defeat and philosophical. How many of us, however, can honestly claim not to be anxious to shoot well? Perhaps we fool ourselves: the ego lurks even where we deny it.

Even Lord Ripon, the best shot of his day (the best shot in England for 40 successive years according to a contemporary) reflected on the gentler pleasures of shooting for its own sake, away from the scrutiny of others:

'...my memory carries me back to a time many years ago when we worked harder for our sport....and I am inclined to think that those were better and healthier days.'

The 2nd Marquis of Ripon, credited as the best game shot of the Edwardian era and perhaps the first 'shooting athlete'. He was a competitive shot of the old school, though he liked to pretend that he did not have to try too hard. Although living to see the perfection of the hammerless gun, Ripon continued to use hammer guns until his death.

However, more typical of Ripon is the reported claim he made of having shot 124 birds for 127 shots, a fact disputed by his loader (Sir John Willoughby's servant, according to Jonathan Ruffer). Ripon could shoot better than anyone, but even he felt inclined to embellish his prowess with the gun. It speaks volumes for the competitive environment that he and his fellows inhabited.

But the practical eccentric, in whose group I classify myself, is fundamentally of the school that sees shooting as a journey, an art form to be developed slowly, without the need for success in competitive terms, though that is not to say we wish to shoot badly. If there is competition it is invariably with oneself and one's quarry. I recently shot a day of driven partridges using a different gun for each drive – and three of them were hammer guns. Why? To see if I could do it, for the pleasure of using the guns, for the personal challenge the feat posed, for the simple joy of the exercise. Had I stuck to my hammerless Purdey, I might have killed more birds: but that is hardly the point is it?

Of course, we do not want to shoot badly but we see the primary challenge as being between the game birds and ourselves on the day. If that challenge can be enhanced by the addition of an interesting old gun, then so much the better. If there is pleasure in taking a favorite sidelock to the pigeon hide and shooting well, how much more pleasure is there in taking an obsolete hammer gun with non-rebounding locks and filling the bag with that wily quarry?

If the object of shooting were simply to kill as much as possible we should all shoot our pigeons on the ground when we go decoying; we should all pick the nearest, rather than the highest, pheasant during the drive; and we should all use heavy guns with 36-gram loads wherever and whatever we shoot. We choose not to do these things because we are sportsmen. The practical eccentric perhaps takes his sportsmanship to another level but it is an extension of the same principle.

Competitions for vintage guns

As interest in vintage guns is re-kindled, so their owners desire more and more to put them to use. This has led to some dedicated shooting competitions for the user of side-by-side guns and it is becoming common for these competitions to include a hammer gun class.

Such competitions do include regular competitive

shooters, more used to over-and-under guns, making the switch to side-by-side for the day. But they also attract a good number of 'game shots' who do not enter competitions on a regular basis yet wield their traditional guns in excellent style.

Two major championships are currently held in the UK: the British Side-by-Side Championship, at Atkin, Grant & Lang's Broomhills Shooting Ground and the European Side-by-Side Championship, held at the shooting grounds of E.J. Churchill at West Wycombe. These competitions are not exclusively for vintage guns, as any side-by-side is permissible. However, they do allow the

Mike Yardley with Ken Duglan of Atkin, Grant & Lang and the 2004 British Side-by-Side Championship trophies in three classes.

Victorian pigeon shooting competitions were big news in the UK and the USA. This trophy commemorates W.B. Bingham beating the famous American Shot, Captain Brewer, in Harpenden in 1888.

Mike's Championship-winning Lang. Note the Jones under-lever and non-rebounding locks; yet this gun out-performed all competitors in all classes despite these obsolete features.

THE PRICE OF 'BEST' SHOTGUNS THROUGH THE DECADES

You can't afford a new Purdey now – but could your father or grandfather?

The price of a 'best' double gun from Purdey (plus VAT where applicable)

1900	1920	1930	1940	1950	1960	1970	1978	1983	2000	2005
$448	$340	$631	$575	$1,050	$1,834	$3,597	$14,395	$16,670	$51,515	$85,454

vintage gun user to compete with users of modern guns over a sporting clays layout.

Mike Yardley struck a notable blow for the practical eccentric (and I'm sure he won't feel insulted by being labelled as such) by winning the 2004 British Side-by-Side Championship, against all comers, using a Lang hammer 12-bore with non-rebounding locks, made in the 1870s. This achievement complemented his performance in the Essex County Championship in the same year, in which he used his 1880s W&C Scott sidelock to beat over 200 over-and-under wielding competitors with a winning score of 96 out of 100 over a difficult course in foul conditions.

There is a well-established event in the USA known as the 'Vintagers' in which participants dress as Edwardians and shoot old guns over several days, including some hard-fought trophy competitions. The event is attended by participants from all over the country and from abroad and includes a number of themed social functions as well as the shooting.

What is a 'vintage' gun?

While writing on the subject and espousing a particular 'breed' of firearm it is probably useful to define exactly what a term as imprecise as 'vintage' is intended to mean.

The main argument I hope to put forward is that we can and should be making everyday use of old guns of all types in a modern shooting context. This means, by definition, that such guns should be usable and practical for the purposes of inland duck and goose shooting, driven pheasant and partridge shooting, pigeon shooting, vermin and rough shooting, ferreting, foreshore wild-fowling and sporting clay practice or competition (though this last consideration is a minor one).

The group of guns referred to here collectively by the term 'vintage' begins with the percussion cap muzzle-loader (the earliest guns approaching practical modern application), and ends with those resembling modern firearms in design but made in Britain before 1960. By this time the Gun Trade was firmly in decline and the era of great invention in the design of sporting guns had long been over.

Most people who contemplate using old shotguns in the field, will be inclined towards the breech loading centre-fire shotgun, either of hammer or hammerless design made in the years from 1861 (when Daw patented

An early non-rebounding lock hammer gun by Charles Lancaster. Affordable, beautifully crafted and eminently shootable. Note the bar action sidelocks, hammer bolts and Damascus barrels.

his centre-fire cartridge) to 1890 (by which time the side-by-side hammerless ejector that we know today had become the norm). The Great War struck British society a heavy blow and the gun industry felt the weight of this blow. Many of its next generation of customers were 'lying in a corner of some foreign field' by 1918 and the industry struggled to recover. Many believe the beginning of WW2 heralded the end of the era of the best of British gunmaking. Many skilled workers went to fight and never returned. Many that did return abandoned gunmaking in favour of better-paid industries. Certainly, life after the war was never the same and the British gunmaking industry went into serious decline.

This book concentrates on British guns made before WW2, for here we have the true alternative to the guns in current production. The second-hand market provides us with many variants on the types of gun with which most shooters are vaguely familiar. Here then is our

subject matter. For I believe it to be true that vintage guns offer the modern sportsman better quality, better value for money, better performance and more pleasure than can be had from a new gun.

Why do most people shoot with a machine-made over-and-under?

The norm in the UK for the double gun was the side-by-side – from around 1820 (when percussion cap ignition was perfected) until the early 1980s. From that point in time, when the final vestiges of prejudice against the over-and-under seemed to vanish from the formal game shoot, to the present day, most of the guns that one encounters at any kind of shoot are over-and-unders. This is of no consequence. Barrel configuration is a matter of personal preference and many of the vintage years and finest makers featured unconventional designs such as Lancaster's four barrelled guns and the three-barrelled offerings of Boss, Dickson, Green and Westley Richards. They, however, never caught on in the way that the over-and-under did.

As well as being over-and-under in configuration, most guns we see today are of the machine-made, mass-produced variety and I believe this is noteworthy. What has led the shooting public to abandon the traditional style of shotgun in favour of these largely bland, admittedly functional, chunks of alloy?

History has a part to play and the history of the decline of the once pre-eminent British gun trade is well-documented. The great factories of Birmingham vanished as the world changed and old craftsmen were replaced by machinery. Guns from Spain began to arrive in the UK – copies of English guns sold at subsidized prices the English could not match. One by one the old firms merged and merged again until they went out of business. Only a few remained, largely at the top end of the trade, producing very expensive hand-built guns for the wealthy few. The bulk of the trade died, as did textile firms and, later, steel works and ship builders, as the cost of production in the UK made the products uneconomical in the face of cheap overseas competition.

Now, if you walk into your local gun shop intending buy a new gun that will not cost more than your annual salary, the names you are faced with are not Webley & Scott or Charles Osbourne of Birmingham or Horsley of York, Gibbs of Bristol or Green of Cheltenham. In their place are Beretta, Browning, Laurona, Kemen, Yildiz, Baikal, Arietta, Famars, Krieghoff and any number of European names producing shotguns in all grades from a few hundred pounds to many thousands. You will still find some British names, such as MacNab and William Powell in your price range but these are largely Spanish or Italian made and English badged for marketing purposes. Times have changed.

My father was given a new Webley and Scott 700 series boxlock ejector at the age of eighteen, in 1959. No equivalent English gun exists today. People have become used to the machine-made over-and-under to the point that in a little over 30 years we have lost sight of the fact that English guns are even an option. The European or Japanese over-and-under is the natural choice. It has become so dominant that the over-and-under is the only barrel configuration that many sportsmen under the age of 40 have ever used. Many, if not most, would consider nothing else.

The author's Webley & Scott 700 bought new in the late 1950s for his father. Sadly the firm too became a dead duck in the 1980s.

The practical case for buying vintage guns to use

I will outline a number of key issues one should consider when buying a gun for shooting game or vermin, evaluating the benefits of the new guns available or a second-hand vintage English equivalent. The price range I will focus on is that within reach of most customers seeking a quality, usable shooting companion rather than the wares of the very top makers.

Key factors in the evaluation of a prospective purchase:

1. INITIAL COST
2. RESALE VALUE
3. SERVICING AND MAINTAINANCE COSTS
4. DURABILITY & RELIABILITY
5. QUALITY
6. AVAILABILITY OF SPARES
7. CUSTOMIZING OPTIONS
8. FITNESS FOR PURPOSE
9. AESTHETICS
10. SAFETY

Cost comparisons

The cost of buying anything is at its highest when the item is new. A new Purdey sidelock ejector will cost more than a second-hand one, just as a new Jaguar will cost more than a second-hand one.

In fact you can buy a top quality second-hand Boss, Holland & Holland or Purdey for much less than you will pay for not only a new model from these makers, but for less than you would pay for the best offerings from Famars, Krieghoff, Arietta, Browning or many other current makers.

The key factor for consideration is that the new gun by a British maker offers you no features the old gun does not. The guns are the same; the only difference is a few years. Since it is acknowledged that the best output of the British gun industry is the best in the world, why would you want anything else?

In short, the buyer with $20,000 in hand can afford a vintage gun of the first quality but can only afford a new gun of lesser quality. On a cost basis, you get more for your money second-hand. The table below shows a comparison of guns offered for sale in November 2004.

In summary, in each of the categories shown, the second-hand gun is a better quality gun than the new one of equivalent price. It will be better in a year, even better in five years and better still in thirty years.

THE PRICE OF NEW AND SECOND-HAND GUNS

Range $	New Guns $	Second-hand Guns
20,000	AYA No1 Deluxe SLE (21,750)	Holland & Holland 'Modele Royal De Luxe' SLE (1919)
15,000	Arietta side-by-side SLE (14,800)	James Purdey SLE (1889)
10,000	William Powell 'Heritage' SLE (9,900)	Henry Atkin 'The Raleigh' SLE (1931)
6,000	McNab 'Highlander' (o/u) (7,400)	Stephen Grant SLE (1886)
4,000	Beretta Gold E 682 (o/u) (3,900)	Webley & Scott Model 702 BLE (1964)
4,000	McNab 'Woodcock' SLE (3,900)	Stephen Grant hammer gun (1867)
2,000	McNab 'Lowlander' (o/u) (1,570)	Henry Atkin SLE (1900)
1,000	Yildiz side-by-side .410 BLE (990)	W.R Pape BLE (1914)

SLE – Sidelock Ejector

BLE – Boxlock Ejector

All guns are 12-bore unless otherwise stated.

The 2007 William Powell 'Heritage' sidelock retails for $11,260. It is made on the continent and finished in the UK. An entry-level 'English' side-by-side.

The chart challenges Mike George's assertion in his *Shotgun Handbook* of 1997 that the average shooter with between $1,000 and $10,000 to spend is better served by European, American or Japanese machine-made over-and-under guns. He was only considering *new* guns: a serious oversight. Think about it: do you want an Arietta or do you want a Purdey?

THE **Webley**

HAMMERLESS EJECTOR GUN

Lot 451

KNOWN FOR ITS HIGH QUALITY WORKMANSHIP, PERFORMANCE AND RELIABILITY.

ASK FOR CATALOGUE – WEBLEY & SCOTT Ltd. BIRMINGHAM.

The 1950s entry-level side-by-side – The Webley 700 – made in Birmingham, perhaps it lacks some of the finesse of the Powell but it is available today in top condition for less than $2,000.

Resale value

In 1883 the price of a new 'best' quality Purdey sidelock ejector was $320, in 2000 the same gun cost $60,750 new (plus VAT). As I write I have the December 2004 catalogue for Bonham's auctioneers. It features an 1891 Purdey SLE in excellent condition with a $11,000-$16,500 estimate.

I asked Nick Holt, Director of Holt's auctioneers, who specialize in sporting guns, to estimate the resale value at auction after three years honest use of some of the guns he sold in 2002, assuming they suffer no damage nor develop anything needing rectification. Nick told me he would expect the guns to make as much, if not more than they made when last sold. When you buy a new gun, it will depreciate with each passing year of ownership. This is especially true of the cheaper end of the market and the higher end. The mid-range over-and-unders of Beretta and Browning hold their value rather better.

A friend recently bought a Gamba Daytona, retailing at around $7,900 new, for $3,162. Spend the $7,900 on an English sidelock and your money is safe. In fact, Peter Pedder, who specializes in vintage gun sales, recently offered me a lovely Purdey thumbhole-lever hammer gun for $6,900. He had previously sold the gun in 1995 for $2,370.

It can be seen from this comparison that money spent on a quality vintage gun will be money well spent and that three years of further ownership is irrelevant to the market and its future value (or five years for that matter and probably ten).

In summary, buying a vintage gun is an investment in a piece that will not lose value if cared for and may actually gain in value with the passing years. New guns by contrast will lose money the moment they leave the shop and continue to do so for some time.

Servicing & maintenance costs

Well-made guns need very little maintenance and well-made guns that have been in service for a hundred years or more will not have much likely to go wrong *if they have been cared for*, as the materials from which they are made are excellent and the workmanship that went into them was of the highest quality and carried out to within very fine tolerances. Everything works in harmony with everything else and has had decades to 'bed in.' Apart from the occasional need to have a spring made (every 40 years or so), there is little else to do to maintain a quality vintage gun except clean it and have it checked by a gunsmith every few years.

In fairness to the modern factory-made gun, they are also very reliable tools and the well-known models have been around long enough to have sorted out any 'teething troubles'. An established model from Beretta, Browning, or Perazzi or a gun of similar quality is likely to be fairly bullet proof. The availability of parts is good and the dealer network good enough to sort out most problems in a straightforward manner.

The problems with new guns tend to come with very expensive and very cheap models. For example, a shooting acquaintance of mine, and well-known TV personality, bought a top-of-the-range Italian sidelock for over $20,000 two seasons ago yet is still using a borrowed Beretta because his expensive Italian gun kept firing the two barrels together, or refusing to fire at all. It was sent back to the vendor on a number of occasions but no improvement resulted. He is now trying to get a refund. Had he spent the money on a second-hand Purdey, he would not have had the problem.

At a clay shoot (actually the European Side x Side Championship) in 2004, I observed a young chap using a new Yildiz .410 in the small-bore category. These cheap Turkish guns look nice and appear to be a good deal in the sub $990 bracket but this one would not fire the second barrel. These problems with new guns are only going to get worse. Had the $990 been spent on a solid English boxlock non-ejector, there would have been no sorry story to tell.

Finally, a basic comparison of the cost of servicing each type of gun at the end of the season will be useful. I asked Park Street Guns in Hertfordshire, a well-established

THE COST OF SERVICING GUNS

Gun Type	Cost
Beretta 686 over-and-under (2002)	$170
William Powell side-by-side sidelock ejector (2002)	$355
Beretta side-by-side boxlock ejector (2003)	$280
Webley & Scott 700 series boxlock ejector (1957)	$280
J. Purdey & Sons sidelock ejector (1889)	$355
H.E Pollard boxlock non-ejector (1910)	$275

All prices are exclusive of parts.

high-street gun shop, to quote for stripping, inspection and reassembly of various guns (*see table*):

The trigger plate action Beretta over-and-under is cheaper to strip and clean than the other types. The 1889 sidelock and the 2002 sidelock cost the same and the age of the boxlock also makes no difference to the service costs. In all cases, the routine annual service is trifling in comparison to the servicing of a car or other mechanical apparatus. Age is nothing to be scared of.

Durability & reliability

The low overall servicing cost is generally due to the high-quality workmanship and materials used to build vintage guns. Parts very rarely fail. The designs, though varied and ingenious, have stood the test of time. When a new gun is launched, it is not unusual for the design to require some modification before being proven through hard use.

This is not a problem with new versions of proven designs such as the Beretta 686 series and Browning B25 and others. However, a good case in point is the Browning Cynergy, launched in 2004 and by the summer of that year subject to a recall of all guns sold in order to address a production fault.

Many top brand foreign guns are made well but not as well as the guns made by the top English firms at the turn of the 20th century. Reliability problems are very likely to develop in the medium term with cheap new guns. If your vintage gun has had regular use and good care for fifty years or more, the likelihood of it malfunctioning tomorrow is low. The only thing likely to fail is the occasional spring, through years of honest wear.

A high pheasant dead in the air. The picture, taken in the 1940s, shows classic side-by-sides in action. The guns used here are almost certainly still in use: the longevity of bench-made English shotguns is astonishing.

Quality

As previously stated; if you buy a second-hand car for $20,000 you will get a better quality car then if you buy a new car for $20,000. For example you may get a four- year-old BMW 3-series rather than a brand new Ford Fiesta.

The reason people choose a new car is for considerations of depreciation, worries about reliability and the potentially high costs of service and repair for the older top of the range car.

We have seen that the worries of the car owner, listed above, are not material for the gun owner but the basic purchase principle still works. For $8,000 you can get a 1910 Woodward sidelock ejector in good condition or a new decent-grade Perazzi. For just under $1,000 you can choose between a 1900 Greener boxlock non-ejector pigeon gun with best Damascus barrels and a new Yildiz side-by-side ejector with a gold-colored single trigger.

In each of the above cases the vintage gun is of far higher quality. If new, it would cost thousands of pounds more but it is still as serviceable now as it was fifty years ago – and still will be in another fifty years; long after your new gun has become just another unremarkable second-hand import.

In the matter of quality per pound/dollar spent, the second-hand vintage gun is the clear winner.

To this I should add a caveat. One still needs to be careful when buying old guns; there have been many opportunities over the years for them to be neglected or abused. Ask a gunmaker what is most likely to cause problems: 'Single triggers and ejectors' will be the reply. It was also sagely observed by a friend in the trade that over the years gunsmiths (or gun dealers) have ruined more guns than have been worn out through hard use. Indeed, some of the practices employed over the years to get a gun saleable in the short term and turn a profit make one weep.

Foreign machine-made guns, however, are generally reliable and durable. This is what has made them so popular. A new medium-grade gun in the $4,000-$6,000 bracket will work well for years and be a dependable shooting companion. It will be the practical equal of the robust Birmingham boxlock ejector that it replaced as the backbone of the shooting scene. In terms of reliability and durability, old and new guns tie for the honours.

THE BASIC PRICE OF A BEST SIDE-BY-SIDE 12-BORE SHOTGUN (2005)	
Westley Richards 'hand-detachable boxlock'	$48,300
Holland & Holland 'Royal' SLE	$85,900
A.A. Brown & Sons Supreme De Luxe assisted opener SLE	$58,300
James Purdey & Sons SLE	$84,700
William Evans SLE	$63,750
E.J. Churchill 'Premier' over-and-under	$72,000
Boss & Co SLE	$100,200
Watson Brothers SLE	$51,900

These prices are the lowest possible charged by each firm. Most include standard engraving. Any non-standard features such as single triggers, higher-grade wood, rib or stock variations or side-clips are charged as extras. VAT is not included in the prices – so add 17.5% if it applies.

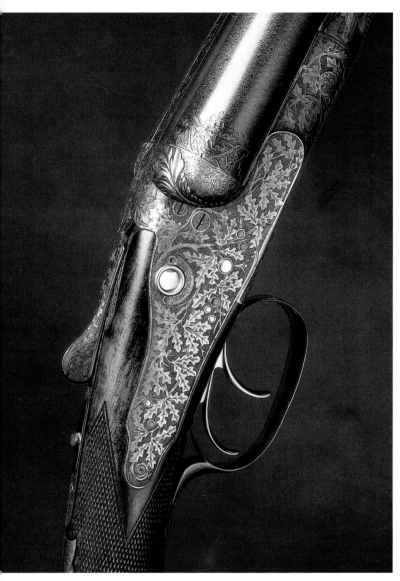

The quality and beauty of vintage guns can be breathtaking. This Scott & Baker patent (1878) back-action pigeon gun shows extra special engraving, figured wood and quality Damascus tubes.

Availability of spares

Here we have the new machine-made gun in the advantageous position of having many interchangeable parts. If you need a new stock, just buy one and bolt it on; barrels are also readily purchased and fit straight on to the action. If you bend the trigger guard on your Beretta, just order one by post; it will arrive and you can screw it on.

High-grade modern guns (of whatever origin), which are hand-finished, and all English vintage guns, will need to have replacement parts made to order. This is clearly more expensive.

However, you will only need a new stock or barrels if you break them through serious misuse or accident. For such eventualities, insurance is required and recommended. In this case the machine-made gun is still at an advantage because the cost of insurance will probably be lower.

In conclusion, the machine-made gun is the easiest and cheapest for which to obtain spares. This is not however, as significant as one would think when considering the infrequent actual need for replacing the parts of a well made, well cared-for gun.

There is no part of a vintage gun of proven design that cannot be replaced by a skilled gunsmith; once replaced the new part is likely to last another half century or longer.

Customizing options

Some people like to adjust any gun they buy to reflect their personal needs or preferences.

Both modern and vintage guns can be altered to suit if so desired. Typically, this will involve adjusting the stock to fit the shooter for bend and cast, adding to or reducing the length of the stock and changing the chokes. There is little else to do to a gun that makes any appreciable difference.

A skilled gunsmith can adjust either a modern or vintage gun to suit in any of the above categories with equal ease. Vintage guns can have choke tubes fitted in the modern style and all the other adjustments are well-established practices of gunsmiths and go back to the dawn of the breech-loader.

In the category of customizing options, there is nothing to choose between the old and new gun. An example of this principle is Jason Abbott's 1920s Purdey pigeon gun. This has been re-stocked to fit the owner with a high 'Trap' style comb and 'Prince of Wales' grip. It has been sleeved and fitted with Teague screw-in chokes and is exactly the gun the owner wants for tackling modern sporting clays. Jason bought the gun at auction with worn-out barrels and a damaged stock. The customizing reflects his needs and desires and it is, in essence, a bespoke gun.

Fitness for purpose

Modern guns come in a variety of guises, semi-autos with camouflage plastic stocks, over-and-unders available as heavy 'trap' guns, 'sporters' for sporting clay shooting and 'game' guns for use in the field. These are just some of the types of gun encountered nowadays and the choice seems bewildering. It is easy to think we have more choice now than we have ever had. However, this is not so.

In the past, gunmakers produced specialist guns of all manner: heavy, straight, strong 'live pigeon' guns, plain, tough, simple wildfowling magnums, lightweight guns built to take light loads, heavy guns designed to take heavy loads, guns with pistol grips, semi pistol grips, straight hand stocks. Guns with different ribs, different actions and different proportions were commonly available. Everything from 2-bore to 9mm shot – in fact the vintage gun gives you a better chance of finding exactly what you want than the new gun because you have over a hundred years of production from which to choose, not just what is currently in production.

Should you want a very light gun for walking up game early in the season, vintage guns offer you much more choice in 12-bore than any current new-production over-and-under, due to the heavier weight of the latter. This partly explains the current fashion for the 20-bore over-and-under as a game gun. The 20-bore sits much more neatly as an over-and-under than does the 12-bore and it allows for a smaller breach and action. In fact, as Mike Yardley points out in *The Shotgun*, a modern 20-bore over-

Competition live pigeon shooting, just as modern clay shooting disciplines, generated a particular style of gun dedicated to the sport. They tend to be heavy, tightly-choked and robustly constructed. Some are very finely finished.

and-under and a vintage 12-bore side-by-side game gun are both likely to weigh in at around 6 ½lbs. The 12-bore, however, handles a 28 or 30-gram load better than a 20-bore and will produce better patterns.

Side-by-side vintage guns balance and handle extremely well and are faster to manipulate in the field, shooting 'gun down', while a heavier over-and-under will be more stable and consistent for 'gun up' clay shooting. It is difficult to account for the apparent 'life' in a well-made vintage side-by-side but those who have never tried one should do so a few times just to see what they are missing.

Should you require a heavy, tightly-choked gun for duck flighting or very tall pheasants, a 'live pigeon' gun may suit the purpose. For a fast-handling 12-bore for snap-shooting, try a Churchill XXV or a Lancaster 'Twelve Twenty' or, if you are more adventurous, a Turner 'Featherweight' hammer gun may be an attractive alternative with light loads. As an all-round game gun, a good 28" barrelled sidelock or boxlock weighing around 6¾lbs will give good service. British makers produced these in a wide variety of grades. The choice is vast.

There is a vintage gun for all purposes with greater variety of size, weight, balance, rib, strength of action, proportion and suitability than available in current production guns. Gunmakers in the past offered a huge variety of options and, as guns were hand-made, any combination of desired features could be fashioned to best effect to produce the finished gun. Modern guns are almost always a compromise because a standard format and action has to be used as the platform for any variant.

A vintage gunmaker would file an action down to the optimum size for the type of gun ordered, barrels would be made to fit the action and furniture made to fit the actual gun proportionately. This way symmetry and handling is optimized. Many foreign guns offer various bore sizes but they are often on the same action, particularly in the smaller bores. It is common for example to see modern 28-bores built on 20-bore actions or 16-bores built on 12-bore actions. It is increasingly the case that every gun is offered with 3" chambers, interchangeable choke tubes and steel shot proof as standard, regardless of its intended use.

With careful examination it becomes apparent that the vintage gun may be selected exactly in accordance with the intended use for which it is purchased. The foreign gun is usually a compromise.

Aesthetics

The finish on vintage gunstocks will usually be of rubbed oil, created by hand polishing the stock with a linseed oil-based mixture over several weeks, until the hardened end result is complete. This finish is weatherproof and durable, as well as long lasting and attractive. If scratched, it can be rubbed with a little oil and be weatherproof once more. Modern guns are usually varnished or treated with a synthetic sealant. When scratched, they require totally refinishing in order to restore their looks. In time they look progressively worse, whereas the vintage gunstock will maintain its looks and weather gracefully as the years go by.

The stock shape on vintage guns varies and is hand carved to suit its original purpose. They tend to be finer in appearance than the mass produced, heavy- featured blocks attached to the average modern 'off-the-peg' gun.

The finish on the metal parts of vintage guns of quality will be blacking (or blueing) of the furniture and barrels (browning if Damascus barrels are used). Old fashioned blacking is resilient and hard wearing, as well as uniform and fine in appearance. Modern chemical blacking on machine-made guns has a thick, uneven appearance by comparison.

Hand-made vintage barrels are 'struck up' by hand and are shaped to provide weight and strength where it is needed; excess metal is removed elsewhere. This gives the

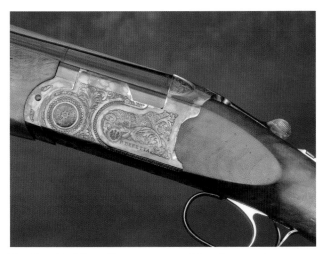

The Beretta Silver Pigeon – an outstanding modern gun at a sensible price.

barrels more grace, balance and style and ensures that surfaces are smooth and polished, bringing out the beauty of the browning or blacking.

Vintage guns are hand-engraved, each one unique and beautiful in a different way. The engraver will have done his best work for the price paid and the engraving will match the shape and style of the gun. Machine-made guns have no engraving worthy of the name on their cast alloy actions. What there is will be uniform and dull, varying very little from model to model. There is no artistry in it, no personality. Every vintage gun has plenty of personality, even a relatively plain one, as long as it began life as a gun of reasonable grade.

Above: The 1909 Boss action (Robertson's patent) is seen here on a lightweight 12-bore o/u with 25" barrels, weighing under 6lbs.

Right: 1913 Woodward patent, these two guns are acknowledged as the most successful and beautiful of the British over-and-unders.

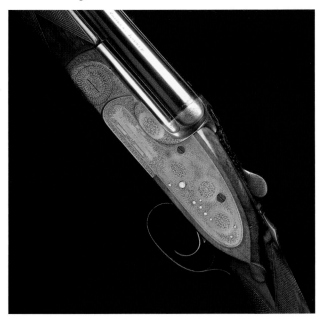

Safety

Users of guns have a responsibility to be rigorously safety-conscious at all times. In viewing an old gun that one is likely to shoot, safety concerns will focus upon two areas: condition and design.

If the gun is in poor condition through wear, it may indeed be unfit for use. The buyer of vintage guns needs to become adept at assessing the condition of the barrels, the fit of barrels to action at the breech and the wear on the bents and sears and other areas of the lock-work. However, this deficiency cannot be solely levelled at the vintage gun because it also applies to any modern gun bought second-hand. One never knows what a previous owner may have done or how well he cared for the gun.

It is allowable to argue that poor quality guns of age may be unsafe due to the workmanship, materials and wear and tear. However, buying a vintage gun of quality offers little risk, as most were over-engineered. A firm's fortunes rested upon a reputation for safety – if word spread that guns accidentally discharged or failed in normal use, it would be a disaster in trading terms and the gunmakers of old were too canny to allow this to happen. Henry Sharp, writing in '*Modern Sporting Gunnery*' in 1909, offered the reader this advice on buying lower-priced guns: '*If you want a best gun, go to a best maker. If you need a medium or low grade gun, go also to a best maker*' precisely for these reasons.

As long as a vintage gun is 'in proof' and in sound condition (and both these matters can be checked with ease), it will be equally as capable of hard use as a modern gun.

Safety procedures for using hammer guns are distinct and need to be learned but they are no more dangerous than hammerless guns if properly used. The addition of an intercepting safety device to prevent the tumblers falling unless the trigger is pulled will be found in some better quality vintage hammerless guns, as well as some modern ones. It is equally likely to be omitted in

Above: The modern Purdey sidelock game gun. A 1900 version is equally beautiful and functional but can be had today at affordable prices.

vintage and modern guns and is certainly not universally applied to guns produced today. For most, the safety catch is no more than a trigger-locking device.

In matters of safety, modern hammerless guns offer nothing that cannot be found in a vintage gun – hammer or hammerless. The crucial factor regarding safety is the user. Whatever your gun, never allow the muzzles to point anywhere you are not prepared for them to discharge.

Above: Better guns like this Purdey sidelock have intercepting safety sears but most modern machine-made guns have saftey slides that engage nothing more than a trigger lock.

The author discussing a 1921 Holland & Holland 'Royal' with a Norwegian firearms dealer at Holt's in March 2007. The gun sold for $12,850 and turned up in a dealer's showroom in the UK a month later priced at $26,700. It pays to know how to sort the wheat from the chaff!

PART II

VINTAGE GUNS AND HOW TO EVALUATE THEM

'Guns are fascinating things, but they are only a means to an end, and to become too wrapped up in them is looking at a grand sport – almost a way of life – through the small end of the telescope.'

Gough Thomas, *Shooting Times*

As a self-confessed lover of vintage guns, I must take issue with 'GT' on this. There is pleasure to be had in the things themselves – as objects of historical interest, examples of inventive engineering wonder, sculptures in wood and metal and as tools for the job. Being a keen Shot and being an aficionado of the gunmaker's art are not mutually exclusive positions to occupy. However, I know where 'GT' was coming from – the acquisition of never-to-be-used 'coffee table guns' interests me not a jot. The much-heralded 'commemorative Christmas cakes' in gunmaking that we see from time to time are not fare for the practical eccentric. Just as well – for we cannot afford them. I like to think we have the taste not to lament this state of affairs.

The collection of the Practical Eccentric

Some collectors have a very specific idea when they begin a collection. It may be that a wealthy individual has a Boss 12-bore and decides he will collect Boss guns. He then seeks one in each bore size until he has a 'full set'. For some this will be a set of vintage guns, for others each one will be a new gun. Another Boss aficionado may want to get hold of as many different types of Boss gun that he can find to form a collection illustrating the history of the firm's production over the generations.

Some of the finest collections have been those connected with the development of the sporting gun, showing all the major changes leading up to the present day. I have heard of other collections that consist of only guns bearing the name of the man who made them, rather than the retailer.

Clearly all the collections listed above are very expensive propositions and the motivation is arguably more about acquisitiveness than it is about the guns, although by no means is this always true. So, what is your collection to contain? Is it to be a private museum or a sporting battery

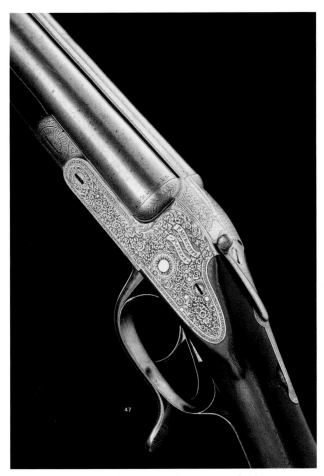

Murcott's 'mousetrap' was the first successful hammer-less gun, patented in April 1871. The under-lever acts both to withdraw the barrel-locking catches and cock the tumblers.

or both? 'Both' being a perfectly reasonable goal of the vintage gun collector.

There are two good reasons for buying old guns: to use them and to enjoy their beauty and their history. We shall leave 'investment' aside for here we are interested in an eccentric hobby rather than a business plan: although if bought wisely, vintage guns will not lose any of their value and may well appreciate.

Guns for aesthetic and historical interest

I am subconsciously thinking to myself as I write this that my own definition of such a gun as suggested by the

Without such a certificate guns will have to be damaged by deactivation or may not be legally sold in Britain without passing proof. If your gun is not deactivated or sold with a certificate of unprovability, you will have to keep it in the gun safe with all the guns in working order, and will not be able to hang it on the wall. Generally guns that do not meet the requirements of the proof house will be sold at auction as 'stock, action and forend only'. The barrels will not be released, except for sleeving, after they have been cut.

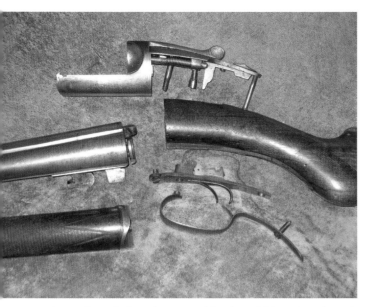

Even cheaply-bought pieces like this 1882 T. Woodward patent spiral-spring 12-bore can bring a great deal of interest to a budget collection. This possibly unique old gun cost only $65 at a provincial auction.

title is *'a gun that does not work'*. This is a purchase for the aesthetics or the history. For why else would you buy such a gun?

My own collection contains few of these. One is an old air rifle, another a cut-away demonstration piece, and that is the end of it. I almost always buy guns I can find a use for. When money is in limited supply, I tend to think it should be spent on a 'shooter' rather than a 'looker'. If I do buy a gun that does not work, it tends to be because I plan to make it work in future.

However, a collection is not just made up of complete guns, it also contains bits of them. I keep an eye open for boxes of old gun-locks, hammers, forends, stocks, actions and other such things. They can become the subject of a lot of investigation and discovery and in many cases can clean up to make attractive display objects.

Should you plan to buy a gun in the UK that does not work, make sure that it has a de-activation certificate or a certificate of unprovability. The latter are now much harder to obtain than was the case. The proof houses have decided to take a harder line of late. A certificate of unprovability was usually granted to an obsolete historical gun that was unlikely to stand proof and was not intended for use when sold. This enabled a UK trade in such guns to continue.

Owning vintage rifles on a Shotgun Certificate

British firearms legislation makes ownership of rifles far more difficult than ownership of shotguns. Clearly, if the object in question is indeed a rifle, it cannot be owned on a Shotgun Certificate. The relevant definition of a shotgun in British law is a smoothbore gun with a barrel length in excess of 24 inches (semi-autos and pump-actions are a little more complicated but not an issue here).

Military Rifles

One encounters from time to time guns that began life as rifles but are now shotguns. I inherited a WW2 Lee Enfield .303 service rifle, which had been bored out to .410 and was now classed as a shotgun. It was totally useless as a sporting gun but quite fun to own as a curiosity.

Double Rifles

Another common conversion is the double rifle into a small-bore shotgun. I recently saw a Holland & Holland .450 hammer rifle, bored out to 28-bore shotgun. This seems a shame – as a rifle it would be much nicer and more use and I began to wonder if it could be re-converted by having 'Paradox' style rifled chokes inserted and used with slug or ball ammunition for wild boar shooting. As a shotgun, these conversions are not very successful, as they are rather barrel-heavy and lack the handling qualities of a game gun.

Rook Rifles

Breech-loading single barrel rook rifles (sometimes termed rabbit rifles) were produced in fair quantities by all the major gunmaking firms (indeed Holland & Holland's sales in the 1870s were over 50% rook rifles) and were generally small calibre (.360, .300, .297/250), single shot designs of various types, though doubles were made. Falling-block actions and hammer guns are common, and Henry Holland patented a hammerless, lever-cocking design in 1882 that was made in large numbers.

They will be encountered with top-levers, under-levers, side-levers and others. Ammunition for many of the calibres used for such guns is now unobtainable and over the years conversions have been performed to bring them back into use. I have known of successful conversions to .22 Hornet and .410 smoothbore. The .410 conversion makes ownership uncomplicated, as the gun can then be owned on a SGC provided the barrel length exceeds 24".

Many of these little rifles were made to a very high standard and make an interesting addition to any collection, as well as being serviceable 'garden guns' for use on rats and other interlopers. In their time they were high-quality, precision weapons, designed to be very accurate to 50 yards or so but without the danger of excessive travel beyond that range.

Collecting guns for shooting

There is a very sensible school of thought that says a shooter should use only one gun. It will make your shooting more consistent and put more game in the bag than if you are constantly changing guns. Clarrie Wilson, the 1960s Olympic coach told Gough Thomas *'Stick to one gun and don't on any account handle any others'*. In terms of shooting consistency it is hard to disagree that the motor-sensory process involved in mounting a gun and swinging on to a bird will be better and easier honed if the gun used is always of the same dimensions and of familiar balance and feel.

This Akrill double rifle was converted to 28-bore shotgun at a later date. Some enthusiasts are reviving obsolete rifles by making their own ammunition.

Note how top snooker players get attached to one cue and their game suffers if they switch. When top Shots had pairs of guns made, the guns had to be matched as exactly as possible to one another. I have even heard of one order for a pair of guns that stipulated Purdey concealed the serial numbers and numbering so that it was impossible to tell if the No.1 or the No.2 gun were being used.

This is where one has to make allowances for foibles. If becoming an Olympic champion, or even a top competitor, is your aim, get one gun and stick to it. If you shoot in company and are concerned that your performance is paramount, stick to one gun. But if you get pleasure out of using the old guns you have because there is sufficient pleasure in that activity alone; then don't hide them away in the gun safe, demoted to the role of useless objects of curiosity: take them out from time to time and give them an airing.

You may find you can adjust more easily if your guns are significantly different rather than all very similar. The brain can make the adjustment needed to shoot your light hammer small bore and your heavy boxlock pigeon gun better than it can adjust between your 6lb 5oz Holland & Holland sidelock and your 6lb 10oz Purdey sidelock. You will just start to find you habitually leave one in the safe and always feel more confident with the other.

I shoot more consistently with my Purdey sidelock than I do with anything else. It fits me, I know it like an old friend and it swings predictably. If I miss, I know every time that it was down to me and not the gun. However, I regularly use other guns just because I like to do so. I

An 1870s Thompson 12-bore hammer gun, as used by the author for driven pheasant shooting in 2006.

especially enjoy shooting guns that had every right to believe themselves to have retired gracefully many years ago. The capacity to adjust is one that humans have in abundance and while it is true that it adds another layer to the considerable challenges inherent in shooting well and consistently, it is one I accept happily.

The essence of the sport of shooting is not filling the bag with as many forgettable birds as possible; the pleasure of the true sportsman is more intangible. I

remember the first pigeon I shot at a considerable range with my old Adams hammer gun on its first outing in the pigeon hide. How pleased was I that it still had the power to do its job after so many years and that I was learning to wield it to good effect?

I remember a fast and challenging partridge stand on a very windy day in Hertfordshire a few seasons ago, not because I shot a lot of very testing birds and was flattered by the compliments of the picker-up standing 200 yards behind me, but because this was the first time my ancient Holland & Holland 16-bore hammer gun and I really 'clicked'. Had I been using my Purdey, the experience would have been less memorable, less special.

There is great pleasure to be found in the use of curious old guns – even in conditions where modern sportsmen have become used to believing that the latest offering from the big gun factories is essential for success. Our forefathers were demanding consumers and the guns they used are still more than capable of performing the tasks for which they were made.

Sleeving

Greener's catalogue of 1888 states: 'Extra barrels usually cost half the original price of the Gun, but no barrels can be properly fitted for less than $51. New barrels for old guns on same terms.'

For many owners, the quality gun with worn out barrels was a real dilemma. Many could not, and many now cannot, afford the price of a new set of barrels. Purdey will happily provide a replacement pair for your gun today – at a cost of around $19,800. 'Unbranded' barrels may be fitted for around $7,900.

What is sleeving?

Sleeving is a process that provides owners of vintage guns with barrels that are too worn or damaged for use, with a means by which to renovate them and use them once more. Geoffrey Boothroyd contends that the originator of modern sleeving was a British self-taught engineer by the name of Ashthorpe, who was repairing guns by fitting replacement tubes to old breech-blocks from 1957 (Gough Thomas cites 1955 as a starting point).

However, the process is not a new idea. A patent, which is essentially sleeving, but intended for original gun manufacture, was taken out by W.H. Monks of Chester in 1881. This states that 'The barrels...are fitted into a breech piece of solid steel on which the lumps and ribs are formed'. This is similar to the modern manufacturing process adopted by many firms for fitting the barrels of new guns, known as 'monobloc'. It is widely used by Beretta, for example in modern gun production.

Vintage British and modern 'best' side-by-side gun barrels are not made this way, they are bored as tubes and fitted to the action by the integral lumps extending downwards, usually by means of a Purdey bolt. Each tube is of one-piece construction. Effectively, sleeving leaves the gun as if it had a new set of monobloc barrels. The jointing seam is often visible to the eye, especially on guns sleeved cheaply in the 1960s and 70s but it can be made invisible nowadays. It is not uncommon for the seam to be engraved to disguise it somewhat.

John Foster Gunmakers sleeve guns using the original system illustrated above; a new tube is mated to the old chamber. The original rib is usually re-used.

Quality

Guns that have been sleeved cheaply and inexpertly will not retain their original balance, dimensions or appearance. If sleeving is done well, however, the sleeved barrels can perform as well as, and look identical to, the originals. However, such a job will be expensive and few guns are encountered with sleeving of this quality. The experience of a generation is that sleeving delivers perfectly safe, robust results. However, invisible and perfect sleeving jobs are not encountered as often as one would wish – some mark of the process is usually visible and it does spoil the aesthetics of a best gun. One gunsmith I know refuses to take on customers' guns for sleeving as he believes the end result is very rarely satisfactory. He recommends owners to keep the original barrels and have a second set made for regular use or that the internal lining process is used.

Damascus tubes and internal liners

Until recently, the sleeving of Damascus tubes resulted in the loss of the beautifully intricate patterns, as the new tubes were steel and the joint either looked odd or the barrels were blacked to look as if the whole were steel. Now it is possible to have a Damascus tube saved by boring out the inside to very fine dimensions (leaving about 10 thou) and inserting a whole new liner, chamber and all. This permits the original weight and balance, as well as the external appearance, of the gun to be retained. The liner can be bored to the original dimensions stamped by the proof house. Guns sleeved this way require very close inspection before it can be deduced that they are not original.

This barrel-lining process has been pioneered in the UK by Nigel Teague of Teague Engineering, At $2,965 for a pair of barrels, it is not cheap but it does offer owners of quality guns with pitted Damascus barrels a means by which to rescue them and restore them to use sympathetically. Because it retains the original appearance of the barrels, it is likely that guns renovated with the Teague sleeve will retain their value better than those sleeved in the original manner.

One very occasionally encounters a gun with barrels that are Damascus at the breech, steel to the chokes and then Damascus again to the muzzles. When done well, as it has been in all the examples I have seen, it is impressive to behold and unusually attractive.

The Teague internal-liner offers a 21st century alternative to sleeving. Shown here is an 1858 Purdey.

This hammer gun has had steel sleeves fitted to the original Damascus chambers and the choke section reverts to Damascus for the final 4" to the muzzles (not shown in this picture).

Proof

Sleeved guns must be re-proofed and stamped 'SLEEVED' (usually out of sight on the barrel flats but sometimes on the tubes, which is unsightly). Those submitted for Teague internal-lining will be stamped in capitals 'LINED'.

Value

Sleeved guns are worth less in the market than those with original barrels. This prejudice can be the basis of a bargain if you want to buy a 'best' gun for shooting and lack the funds. A well-sleeved gun will serve you just as well and will cost you less. A top-quality sleeving job with tig-welded joints will cost $1,680. You get what you pay for in sleeving, as in everything else. Owners who have invested in an expensive sleeving job will end up with a nicely finished and balanced gun. Those who have had it done 'on the cheap' will not.

What to collect

Really there is no answer. Most of us will be constrained by budget; otherwise collecting would simply be a matter of buying everything one saw that one liked. In this respect, I feel there is certain sweetness to be derived from having to cope with the challenges of finding interesting guns at prices we can afford, searching out bargains and looking hard to find what is available that we can find the funds to purchase.

I like to have a use for the guns I buy. I made the error once of buying a 20-bore because I did not have one and thought I should. I have used it twice, I don't like it and it stays in the gun safe. Since then, I have had to envisage using any gun I get for a clear purpose. If I have two that are very similar, one is likely to find itself neglected.

What shooting do you do? What gun would you enjoy using in each case? Don't buy a tightly choked 10-bore boxlock duck gun if you only shoot walked up pheasants and if you generally shoot driven birds, a hammer gun with Jones under-lever and non-rebounding locks is (probably) not for you, though I regularly use one.

I think variety is what makes an interesting collection. I do not want lots of similar boxlocks or sidelocks in mine; I want a wide choice of guns that all offer me a different shooting experience. Other pieces will have value based on curiosity, interest or sentiment. Every collection will be personal and therefore uniquely interesting.

A recent conversation with Mike Yardley led him to reflect on the comment of a mutual friend that Mike instantly lost interest in a gun if it 'did not shoot' regardless of how much it cost, whose name it carried or how attractive it was. These 'one in ten' gems are the ones to look for, the

A unique four-barrel Purdey 20-bore with Damascus barrels. Unfortunately, not part of my collection, but kept at Audley House. The gun lacks the headspace to accommodate modern ammunition and Purdey's have not corrected this to preserve originality. This is a shame as the gun feels lively and begs to be used.

ones that have that inexplicable 'shooting quality' that sets them apart. My $100 1870s Thompson hammer gun has it. Everyone who tried it instantly felt it. They all felt confident with the ugly old nail with the wrong forend bodged on decades ago because it 'shoots'. The 1939 Churchill XXV Premiere I tested recently had it – I felt I could not miss with it the first time I took it out. It sold the next day for $20,000. The chap Mike took it to on approval felt 'it' as soon as he took the gun to the skeet range. These guns one remembers. It is a good motto never to sell a gun that you really shoot well with. They only come along every now and then and often not at a time of your choosing. Hold on to them.

Money is not the key. Some very expensive guns by very prestigious makers do not shoot well. Impossible to explain but true. When you can combine historical interest, original quality and fine aesthetics with a gun that really works, you have a rare thing indeed. The pursuit of these rare things is what gives us few our rare pleasures. Hard graft is the only way to do it. You cannot shortcut the process with money or desire; you need patience, acquired knowledge and a cultivated observation that borders on a sixth sense. It is a pleasant road to travel.

The Twelve Twenty

The 'Twelve Twenty' is a name overwhelmingly associated with the firm of Charles Lancaster of London. It was designed to meet the demands of the public for a lighter weight 12-bore that was faster to swing and easier to carry. Greener, writing on weight ratios, suggested a gun of at least 6lb for firing a 1oz load, based on his 96-1 principle. The 'Twelve Twenty' reduced the weight to 6lbs, or even a little less, and proved extremely popular from the early 1920s.

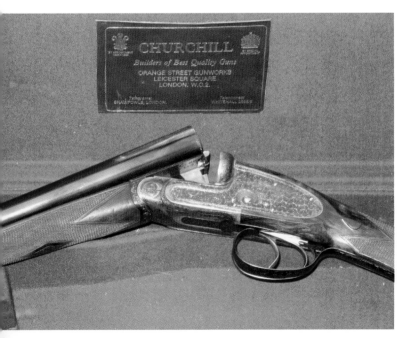

Here the Baker action is seen on a 1930s Churchill 'Premiere' XXV.

It was marketed as having the killing power of a 12-bore but the dimensions of a 20-bore. Catchy marketing was nothing new to the gun trade in the early 20th century (Churchill boasted that his XXV *'Handles like a 20 and shoots as hard as a 10'*).

The design of the 'Twelve Twenty' action was ingenious and provided an exceptional strength-to-weight ratio that enabled the action to be reduced in size without losing strength. It was a sidelock in which the mainspring was housed in a box in the angle of the lock plate rather than being in a recess in the hollowed-out action bar, thus technically an unusual back-action sidelock.

This kept weight between the hands of the shooter, closer to the hinge pin. It produced a gun that was 'lively' and handled very well. External appearance was essentially that of a bar-action sidelock. The tell-tale sign in the picture is the absence of a pin in the forward extremity of the lock plate to retain the main spring.

The tumbler is cocked upon opening, by a separate cocking spring which also pressures the barrel lump and aids the opening action. The mainspring is compressed when the gun is closed. The action therefore works as an 'assisted-opener', which is easier to close than a full 'self-opener' like the Purdey/Beesley. An added advantage is that the springs are at rest when the gun is not being used; though the significance of this is questionable, as many gunsmiths disagree as to whether relieving the springs makes any appreciable difference to their lifespan.

The gun's reputation as an icon of the lightweight game gun enthusiast and the very term '12-20' have entered the psyche of the shooting public and remain alluring in the current climate with the continuing rise in popularity of the 20-bore in particular and light game guns in general.

It will surprise many to learn that the 'Twelve Twenty', so well-marketed by Lancaster as to be synonymous with his name, was actually the invention of William Baker of Birmingham. Baker patented the design in 1906.

Most of the Lancaster 'Twelve Twenty' guns sold were probably made by Baker, as well as invented by him. Other firms also retailed the design under different names: Stephen Grant's 'Lightweight' was one, but Churchill, William Powell and Harrison & Hussey all retailed Baker's gun under their own names.

It is interesting that Burrard wrote '...as this action was first brought out by Lancaster I feel that it (the 12/20) should be honoured with that well-known name

(Lancaster) even though it has been adopted in recent years by …Powell of Birmingham' in his *Modern Shotgun* of 1931, further cementing the 'Lancaster 12/20' myth.

An article on the 12-20 by (the usually reliable) Donald Dallas in the September 2005 issue of the *Shooting Gazette* continues in the same vein and omits to mention Baker, saying '…they (Lancaster's) came up with a unique design…', a curious omission.

Burrard was very complimentary about the design, lauding it for strength, lightness and ease of opening as well as complimenting the crispness of trigger pull as resembling a bar-action rather than a back-action sidelock for feel.

As a modern purchase in the gun dealer's or at auction the vintage 12-20 is expensive and Lancaster branded guns (made between 1924 and 1932) fetch a premium because of the association of the two names and the cachet of Lancaster as a London maker. If you want one, you may be well advised to consider the same action with another retailer's name on the lock if the price differential is significant.

After all, the guns were probably made entirely in Birmingham, many of them by Baker himself and the 'London' and 'Charles Lancaster' association is simply the resilient relic of a long past but far-reaching marketing strategy.

Below: Viewing the guns on sale at Holt's in Chelsea in 2004. Every auction offers you the opportunity to come into close contact with guns of all ages and types.

Buying at auction

Buying guns is something the gun collector clearly has a penchant for doing. Asking any shooter 'What kind of gun do you have?' is a question that generally receives a long answer quoting a list of the various inhabitants of his gun safe: a game gun, a clay gun, a pigeon or rough shooting gun, an old curiosity and probably some other inherited or collected ephemera.

Practical eccentrics like guns and they buy them whenever they can find an excuse. Shooting beautiful old guns that first saw action in seasons long gone and keeping them in active service becomes a pleasure and something of a duty. They are history in your hand and make a shoot day a pleasure; even when there is no shooting. Balance, engraving, patina and various honourable imperfections due to years of wear all add to the experience of using an old gun.

Sources of guns are many; the obvious gun shops, game fairs and friends have been joined as outlets for guns by Internet sites such as guntrader.co.uk and others. However, the auction house has been something of a mystery to the general public for many years. It is a place of uncertainty, no guarantees, no help, and no part-exchange. The guns on sale are also many and varied and problems could be hidden from the untutored eye and lead to the unwitting purchase of an expensive folly.

For too long the Gun Trade has taken their stock from auctions and sold them on to us at a premium because

we were too scared to brave the uncertainties of the auction. I hope to help you understand how the auction scene works and make it accessible. For here, there is fun to be had!

My obsession with old guns has made me a habitual attendee of the London auction scene. I await the arrival of the catalogue from Holt's, Bonham's, Sotheby's and Christie's with eager anticipation and spend days pouring over the descriptions of the coming sale items until viewing day.

There is no better place to become acquainted with the plethora of makers, inventions and styles of guns available to the collector than the auction rooms. It is like going to a museum where you can dismantle and handle the exhibits to your heart's content and then buy the ones you like and can afford, or just walk away a little wiser.

It is worth pausing to reflect on the perception in the Gun Trade that auctions are an increasingly expensive place to buy. It is true that as a buyer you pay a 'Buyer's Premium' of between 15% and 20% of the hammer price, so a gun for which you bid $2,000 will actually cost you $2,400. Add another 17.5% if VAT is to be added to re-imported guns being sold by overseas owners. Sellers will also contribute to the auctioneer's pension fund with a typical 'Seller's Premium' of 10% – 15%. Whatever you do, read the small print in the catalogue before you bid at auction and remember – there is no after-sales service either so *caveat emptor* is the order of the day.

However, auctions can work for you and the excitement is always in finding the bargain of the sale – there are always two or three. There is also immense variety at most auctions – far more than you will encounter in the usual gun shop.

When two or more bidders want a gun, prices rocket (I have even seen quite ordinary used modern guns make more at auction than a new one would cost in a gun dealer's) but when money has been spent, when several similar guns are in the same sale or a gun appears that does not fit the Trade buyers' brief then you can get lucky.

I especially like the auctions for the variety to be found. There is something to suit every pocket and every need. Every interest and niche is catered for and the element of surprise and competition adds to the excitement.

Tips for auction virgins

• Subscribe to the catalogues and read them well in advance, make notes and prepare for the viewing – arriving unprepared and seeing a bewildering array of guns can make your mind go blank. You wander around for an hour and then get home without the key information you need.

• Ignore the guide prices in the catalogues; they are horribly unreliable and usually madly underestimated in order to lure buyers to the sale. Decide what you like and what you are prepared to pay for it. This is the only price of any consequence.

• Don't forget to calculate the cost of 'Buyer's Premium' or you could be in for a shock at paying up time.

• Talk to people at the viewing. 'Gun People' are an eccentric and knowledgeable lot and they love to talk. You can find out a great deal about your potential purchase just by asking the man standing next to you.

I was once looking at a Purdey that had been re-barrelled by the company in the early 1970s and when I was chatting to another viewer about it, he took them off, had a look and said, 'Oh, Alf Harvey made these; good barrel maker'. It turned out I was having a conversation with a man who had been an apprentice at Purdey and stayed there most of his working life!

The author's 1889 Purdey sidelock purchased at auction in 2002. With new barrels by the makers and a beautifully figured Turkish walnut stock of the correct dimensions and proportion, the early Beesley action should see active service for at least another century.

• Go to view the auctions even if you do not buy anything. In fact go to two or three auctions and leave your credit card at home! The more you do this, the greater your intuition for condition and value will be developed.

• Note your top price and stick to it – it is very easy to get carried away and overspend in all the excitement.

• Take your phone and use that useless little camera it contains to snap the proof marks of guns you are interested in for checking out later when you get home – this saves lots of note taking.

• Get a list of the barrel measurements when you arrive at the viewing. Make sure the gun measures well but don't obsess about thick walls if you are buying a gun to use yourself and not to sell on. Dealers look for thick walls (25 thou +) because it makes the gun easier to sell. A quality shotgun with 20 thou in the barrels will still last longer than you if you look after it and could be the basis for a bargain. The right barrel in a side-by-side gets a lot more use than the left and is likely to become more worn over time.

• Be wary of old guns that look very shiny – it is very easy to polish the action and oil the stock to distract the eye from more serious problems. A gun that looks 'tarted-up' probably has been. Look for guns that have honest wear and on which all the bits look like they belong together. Remember that the Trade buys guns at auction but also uses auctions to offload items they are struggling to get rid of.

• Don't underestimate the cost of repairs – a side lock could easily cost over $4,000 plus wood to re-stock well, a box lock could cost $2,000. A new set of barrels for a double can cost between $8,000 and $20,000. Sleeving could be $1,600-$3,000. As you get more knowledgeable, you will become more confident about what you can take on. If in doubt, get quotes before you bid.

• Remember that cheap guns of low quality may be uneconomic to repair. I remember a friend who was partial to renovating old motorcycles telling me 'it costs the same to re-spray a BSA Bantam as it does a Brough Superior'. The same principle applies to guns. Starting a project with a gun of low quality can be an unwise plan of action.

• Finally, remember that if you buy something that you come to regret, you have no recourse to the auction house in the way you have if you buy from a shop. If it has a crack you can't find, tough. If the barrels are dented and you didn't notice, likewise. Until you have confidence in your knowledge – ask someone to check it for you.

• It amazes me how many people buy guns at auction and find they can't shoot with them, then bring them back to be sold again. This is a very expensive way to test a gun before settling on it. If you don't have the skills to assess fit, make sure someone who does checks you out before you bid. Ask the auctioneers for help – there is almost always someone qualified in the room, who can take a quick look as you mount the gun, and they could save you a few hundred pounds.

Guns stacked up in the gunroom of Holt's at Sandringham. Auctions offer a huge choice of vintage guns in one location. Few, if any, gun shops can rival the sheer variety.

What to look for at auction

It is paradoxical that I spend a lot of time studying the catalogues and taking notes before the view dates but upon arrival am often engaged by something that did not appeal on paper. The beauty or usability of a gun may only become apparent when you study it. Other times a good specification on paper is an instinctive and immediate 'no' as soon as you pick it up.

I like to make a shortlist from the catalogue of price ranges and then decide which I am likely to actually bid on. The exercise of evaluating guns in all price ranges is a good one. You may not be in the market for a $30,000 sidelock just yet but one day you may be and you will benefit then from what you learn now.

When I evaluate a gun, I like to predict what it will make and compare this with the hammer price achieved. This makes you good at judging value, helps you spot where the bargains are to be found and makes you a wiser bidder. This is a valuable skill. You learn to ignore the estimates given by the auction house and accurately read the market for guns. When someone shows you a gun for sale elsewhere, you can ask yourself 'What would this fetch at auction?' and give an accurate valuation accordingly.

The trick for the occasional collector and user of old guns when buying at auction is spotting potential in the oddity, the unfashionable, the gun that does not appeal to the trade buyer or the conservative. If 30" barrels are fashionable, buy the bargain priced 25" gun, for as sure as the sun rises every morning, the fashion will turn full circle.

A good example of this kind of thinking was exercised at a recent London auction. A Holland & Holland 'Paradox' hammer gun (a type of shotgun with a rifled section in the choke tube allowing accurate shooting with a single projectile loaded into a cartridge, as well as producing half-decent shot patterns) was for sale.

These are collectable and expensive under normal circumstances but this one had been sleeved (well and by the makers in 1967) and was now a smoothbore without rifled chokes. The bidders overlooked this lot, as it did not fit in with what most people want.

It was no good to top-end collectors because it was not original, not attractive as a game gun as it was heavy and slow to load with its rotary under-lever and no good to the gun shop owner as the average punter coming in would not be interested.

It made $1,100. This was a Holland & Holland! It was fabulous quality, shootable, historically interesting, in excellent condition and it cost no money. Somebody now has an excellent pigeon or wild-fowling gun that will be pleasure to own and use.

Bidding

It is important to know what you want and how much you have to spend but remain alert to the unexpected bargain. You may covet a gun but assume it will be sold for a certain price and concentrate your attention on another. Only

when the bidding is in progress may you become aware that the original subject is likely to go for a bargain price. You need to be ready to react.

Beware, however, not to bid automatically on a lot that you have not studied but which seems to be going for no money. You bid impulsively and end up with a pile of junk that you need to put on your Certificate and find room for in your cabinet. I have been caught out this way on more occasions than I care to tell.

Be sensible, decide what you are prepared to pay for a gun and write it down. Do not exceed this price. Your judgement is sounder in the calm of your study than in the excitement of the auction room. It is all too easy to put your hand up for just a couple more bids, only to find you have paid $2,000 over the odds. You will live to regret this overspend. If a gun exceeds your own valuation and you lose it, despair not. Another will come along in time. Be patient.

The top ten most horrible bodges encountered on vintage shotguns

1. Shortening barrels.

2. Sawing a few inches off the end of a stock to make it fit a youngster.

3. 'Home-made' custom checkering added by a previous owner.

4. Polishing the action to make it shine (but wearing away the engraving).

5. Beavertail forends fitted to 19th century side-locks.

6. Poorly executed sleeving.

7. Lapping bores to remove dents without raising them first.

8. Ejectors ruined through attempts to regulate them by people who do not understand them.

9. Re-application of color hardening (it rarely looks right).

10. Clumsy wooden extensions to the butt.

Nick Holt has expanded rapidly to become one of the major players in British gun auctioneering. In 2005 Holt's were turning over $9,100,000.

Above: Guns on display at Sotheby's in London (2005). Note the chains and missing fore-ends, removed for security reasons.

Left: The auction rooms at Holts in Chelsea (2005). Bids are being taken from the room as well as on the telephone.

Right: Register your paddle, read your catalogue and don't get too excited!

Preparation for the auction

The main auction houses in London generate catalogues of very fine detail and extensive illustration prior to each sale. Subscribe to these and pay your subscription by direct debit, they will then arrive on your doormat about two weeks before the sale.

Here is a typical entry (from the Holt's sale of September 2004):

Lot 603 **J. Woodward & Sons 12-bore Sidelock Ejector**

Serial no.4507, nitro sleeved barrels, rib, re-engraved with makers details, 2½" chambers, bored approx Imp Cyl & ½ Choke, arcaded side-clipped fences removable striker discs, protruding tumbler pivots with gold-inlaid cocking indicators, best fine-scroll engraving, brushed bright finish overall. 14¾" highly figured pistol grip stock with horn pistol grip cap and 1" rubber recoil pad, weight 6lb 10oz. The makers have kindly confirmed that the gun was completed in July 1892.

Now there is a lot of jargon to deal with in that description and wading through seven hundred-odd of these will make your head spin if you do not know your basic terminology. You will certainly find it hard to decide how to shortlist guns you want to examine and note the points to pay special attention to. For example, is a 'protruding tumbler pivot' a good thing, an indication that something is broken or simply a feature of this model of gun that actually makes no difference to its shooting?

To succeed at auction, a basic knowledge of shotgun terminology is essential – it will also help you understand and enjoy your guns better. The range of terms and descriptions of gun parts may seem bewildering at first but it is surprising how quickly one can learn to appraise and describe a gun accurately. It also helps a lot when discussing work with gunsmiths or making notes about condition, features etc. The following pages explain the technical terms you need to understand at an auction:

The barrels

Chambers: Where the cartridges are inserted.

Chopper lumps: Chopper lump barrels are generally a sign of quality and are used for 'best' guns as a standard feature. The barrel lumps are forged with the tubes. When jointed together, a fine line is visible down the middle of the lumps.

Muzzle: The end of the barrels opposite the chambers.

Rib: The metal filling the valley between the two barrels on a side-by-side or on top of the 'over' barrel on an over-and-under. May be found in a variety of shapes and finishes.

Breech: The part of a shotgun where the barrels and action meet. Modern guns open at the breech and are therefore 'breech loaders' as opposed to 'muzzle loaders', which had to be loaded from the only open end (the muzzles).

Extractors: Metal rods fitting into the barrels at the breech. When the gun is opened, they extend forward and lift the cartridges proud of the breech.

Ejectors: Essentially the same as extractors but split so they can act on the fired cartridge only, when engaged by the 'kickers'. Extractors hold the cartridge proud of the barrel for removal by hand, while ejectors throw them clear of the gun. Ejector mechanisms are many and vary from the very simple to the very complex.

Barrel flats: The part of the barrels that sit on the table of the action (generally where proof marks are to be found).

Lumps: Extending from the barrels into the table of the action. In modern side-by-sides the lumps have bites cut into them to receive the Purdey bolt.

Hook: The concave surface on the forward lump that engages with the hinge pin to form a pivot for opening and closing the gun.

Bites: The notches cut into the lumps to receive the Purdey bolt when the gun is closed.

Top extension: One of a number of design variants of extension to the top rib that engage with the top of the action to secure the gun closed, often encountered in conjunction with the Purdey bolt.

Sleeving: A method of renovating worn out barrels by cutting them off ahead of the chambers and mating new tubes into the retained section. Recent developments allow for a complete tube to be inserted into the barrel to strengthen it, while leaving the exterior unchanged.

Loop: The loop is the part of the barrel to which the forend catch engages to secure the forend. It is brazed into place.

Bead: The small brass sight at the muzzle end of a shotgun.

Chokes: Constrictions at the muzzle designed to improve the effective range of a shotgun.

Dents: Damage to barrels caused by knocking against a hard object. Dents can be removed, or 'raised' but doing so can reduce the thickness of the barrels.

Rivels: Swellings in the barrel that cause it to look uneven. Caused by pressure from a small blockage or overloaded cartridges.

'Browned': Damascus barrels were traditionally rusted under controlled conditions to enhance the figure and color in various shades of brown.

'Blued': Or 'blacked' barrels are usually of the modern steel type, although Damascus barrels were occasionally blacked and are occasionally encountered. The trigger guard, lever and other 'furniture' are also usually blued. The terms 'blue' and 'black' in gunmaking are virtually interchangeable.

'Re-blued': Over time, blueing fades and can be re-applied to enhance the appearance of the gun. The results are not always pleasing and barrels that have been constantly re-blued will be weakened and the lettering on the rib will lose its crispness.

Lapping: When a gun is 'pitted' or rusted inside the bores, metal can be removed from the bores by a process known as lapping. This is a common procedure but care must be taken, as the gun will be 'out of proof' if it is lapped out by more than 10-thousandths of an inch. When viewing a gun with apparently insignificant pitting, take careful measurements of the bores: they could be on the edge of proof due to earlier lapping-out of more serious pitting. An unwitting purchaser may think it easy to remove the little pitting evident, only to find the gun is then out of proof.

The locks

Hammer: The hammers are visible on the outside of the locks of a hammer gun. They fall on the strikers to fire the cartridge.

Tumbler: In 'hammerless' guns, the hammers are put on the inside of the lock and called 'tumblers'.

Lock pin: A screw that secures the lock to the action (gunmakers call screws 'pins').

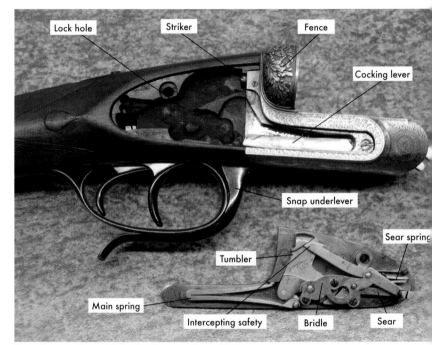

Boss bar-action sidelock ejector circa 1914.

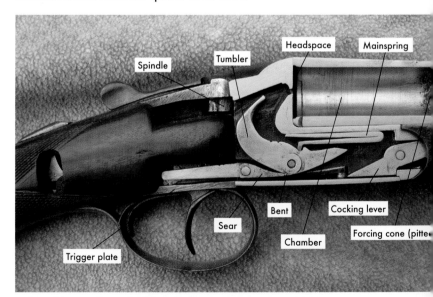

Helson (Anson & Deeley) boxlock non-ejector circa 1900.

Sear (or scear): The sharp point that engages with the bent to hold the tumbler at cock.

Bent: A recess cut into the tumbler into which the sear fits to hold the gun at cock. The quality of the sears and bents is crucial in ensuring the gun is safe and has crisp trigger pulls.

Intercepting sear: Some hammerless actions have a system by which an additional safety feature is the presence of a second sear that engages a bent in the tumbler when the safety catch is 'on'. This makes it impossible for the tumbler to fall unless the safety catch is put in the 'off' position and the trigger is pulled. Otherwise, safety catches are no more than trigger locks.

Cocking indicators: These are sometimes found on the outside of the lock plates of hammerless guns to indicate when the gun is cocked. They were first designed to reassure people who distrusted hammerless guns because, unlike hammer guns, it was not easy to see if it was cocked or not. Some guns have clear windows in the action through which the tumblers can be seen.

Main Spring: The spring that moves the tumblers or hammers. Also used to force open the gun on variants of the Beesley self-opening sidelock.

Side plates: Some trigger-plate or boxlock guns have steel plates attached to make them resemble a sidelock for aesthetic reasons.

Lock plates: Hammer guns and sidelock hammerless guns have the hammers, or tumblers, and springs mounted on plates at the side of the action.

Back action locks: Sidelocks with the mainspring behind the tumbler.

Bar action locks: Sidelocks with the mainspring housed in the bar of the action, which is cut away to accommodate it.

'Round Action': The name commonly given to the trigger-plate gun made by Dickson of Edinburgh. The term is also erroneously used to describe some sidelocks (such as the new Holland & Holland side-by-side – which is actually a back-action) where the bar of the action is filed into a 'rounded' shape.

Trigger-plate locks: where the lock work is housed on the trigger plate, as featured in the famous guns of Dickson and MacNaughton and the modern guns of McKay Brown. Many modern over-and-under guns use the trigger-plate action, the Beretta 686 for example.

The action

Action face: The vertical part of the action that meets the barrels at the breech.

Action flats: The top of the table of the action that faces the barrel flats when the gun is closed.

Bolt: The locking mechanism that holds the breech together. The Purdey bolt, perfected in 1867, has been almost universal since that time.

Knuckle: The extreme end of the action bar, rounded to accept the concave end of the forend, at which point it acts as a hinge.

Striker holes: Holes in the action face through which the strikers emerge to detonate the cartridge when struck by the tumbler. Some guns have threaded discs set into the

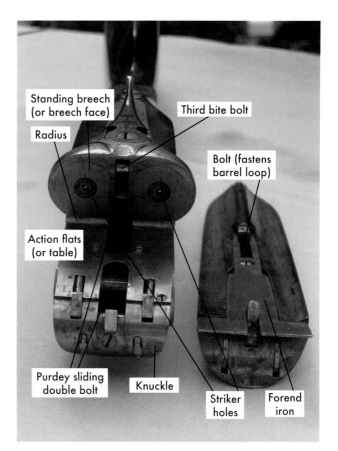

Purdey sidelock ejector (converted) 1889.

action face so the worn striker holes can be easily replaced. Guns so equipped have 'disc-set strikers'.

Top strap: The metal extension of the top of the action that extends along the top of the hand of the stock.

Side clips: Small extensions of the fences that engage with the sides of the barrels to prevent sideways movement. Especially common on guns designed to fire heavy loads such as 'live pigeon' guns.

'Brushed bright': Many old guns are polished when the color hardening has faded and the steel appears dull or tarnished. It is supposed to make them more attractive.

'Self opener': An action in which the gun is forced open by pressure from the mainsprings when the top-lever is rotated. The most famous is the Beesley-Purdey.

'Assisted opener': Essentially the same in principle as the self-opener but not such a powerful operating mechanism, sometimes using only one lock spring.

The woodwork

Pistol grip: A style of grip at the hand of the stock, shaped like the handle of a pistol. It is common on double rifles

Below: Purdey sidelock ejector 1889.

and some foreign guns but not traditional in English game guns.

Semi pistol grip: A more rounded and less angled grip sometimes found on English shotguns, especially pigeon guns and wildfowling guns.

Straight grip: The traditional English gun stock style where the hand of the stock joins the buttstock in a straight line.

'Prince of Wales' grip: This is a less inclined variation of the pistol grip, more elegant and suited to shotgun use – especially guns with single triggers.

Checkering: Lines cut into the hand of the stock and the forend (and occasionally the side panels). Checkering varies in numbers of line per inch, according to quality; generally finer quality guns have finer checkering.

Escutcheon: An inlaid plate of silver or gold, usually an oval or shield intended for the engraving of the owner's initials or crest.

Butt plate: A plate of metal, horn or ebonite on the butt of the stock intended to provide protection when the gun is rested butt down on the ground. Muzzle-loaders had strong metal butt plates but most 'best' breechloaders dispensed with them.

Heel: The top extremity of the butt.

Toe: The bottom extremity of the butt.

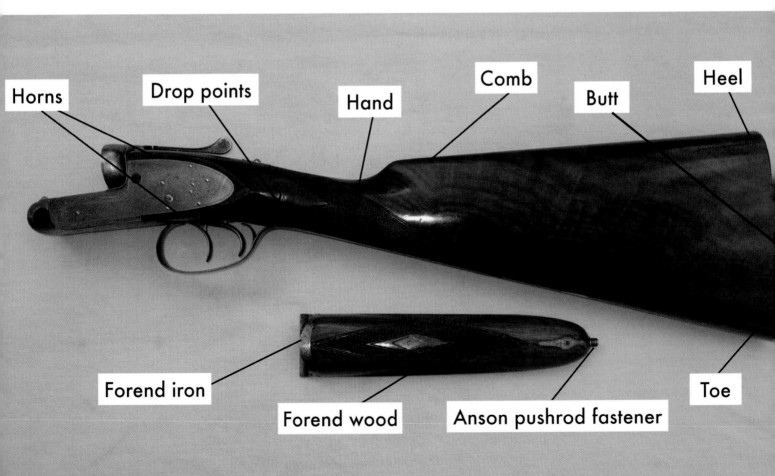

Wrist: Also called the 'hand'; this is the slim part of the stock, which is gripped by the right hand (of a right-handed shooter) when the gun is mounted.

Hand: See 'wrist' above.

Comb: This is the highest part of the stock, on which the shooter's cheek rests when the gun is in the firing position.

Pistol grip cap: A plate of metal, wood or horn covering the flat bottom extremity of the pistol grip.

Oiled finish: English guns traditionally have a finish made from the rubbing in of a mixture of linseed oil and turpentine. This produces a tough, easily repaired surface and enhances the beauty of the wood.

Drop points: Shaped extensions of the side panels, usually in the shape of a water drop. Encountered on better quality sidelocks and boxlocks and some hammer guns.

Horns: The finely cut part of the stock that extends to the action of a sidelock. In best quality sidelocks the horns traditionally extend to the fences.

Triggers and furniture

Single trigger: Employed on double guns in which only one trigger is needed to fire both locks. Two main types are employed; 'three pull' and 'delay' systems.

Double trigger: Traditional double gun configuration employed a front trigger to fire the right barrel and a rear trigger to fire the left.

Articulated trigger: Some front triggers are hinged to allow them to move forward when the rear trigger is used. This prevents shooters with big fingers, or those prone to bruising, from suffering injury when the rear trigger is pulled.

Selective/non-selective single trigger: Some single triggers may be adjusted to change whether the first pull fires the right or the left barrel. In some designs this is effected by a slide on the side panel or top strap.

Side safety/top safety. Most shotguns have the safety catch placed on the top strap but Greener devised a sliding button on the side panel, which he argued required less removal of wood from the hand of the stock, resulting in a stronger gun.

Trigger guard. A metal strap that extends from the underside of the action and which is secured to the hand of the stock to prevent triggers from being pulled unintentionally or catching on clothing or foliage.

Diversions and blind alleys

'There are perhaps some sportsmen who are inclined to treat the repeating shotguns 'au serieux'. The author cannot do so, for although he admits the mechanism to be ingenious, the results obtained by their use do not warrant their general adoption.'   W.W. Greener, *Modern Shot Guns*, 1891.

Greener's blunt pronouncement notwithstanding, the desire for increased firepower was one of the main thrusts throughout shotgun design evolution. From around 1860-1905, guns moved rapidly from being percussion muzzle-loaders, that were very slow to put through the discharge-reload-discharge cycle, to the advent of the semi-auto that could be loaded with five shells and discharge all of them

One of a pair of 3-barrel 20-bore Dickson trigger plate game guns. Mike Yardley has shot several of these and commented 'fun but no advantage.'

in a few seconds. Ejectors were mainly designed to speed up the 'shots per minute' rate of a gun in a hot seat on a driven day. This desire for speed was also a key factor in the decline of the hammer gun.

Despite the plethora of inventions designed to increase the rate of fire, the option that found most favour for driven game shooting and still does to this day is a pair of hammerless ejectors with Gun and loader working as a team. The rate of firepower possible for a skilled duo is phenomenal and the great advantage is the short time required for reloading. A pump action or semi-automatic gun may be able to fire 5 shots in quick succession but it takes longer to reload. The fact that these guns cannot be opened at the breech and made visibly safe also added to their perceived unsuitability for formal shoots.

The re-loading speed of even a single hammerless ejector gave it greater popularity than repeating arms managed to achieve, though for rough shooting and wildfowling, pump-action and semi-automatic guns did find favour in some circles and this branch of shotgun design went in its own, separate, direction.

Before polite society settled on the use of a pair of sidelock doubles, inventors and gunmakers had laid a

This Edwinson Green three-barrelled 12-bore is a trigger plate action of Green's own design, patented in 1902. At 8lb 5oz it is little heavier than many modern over-and-unders. I found it well balanced and pleasant to handle and would love to have tried it in the field. The single trigger fires the right, then the left, then the centre and the barrels are choked Improved Cylinder, ¼ and ½ respectively.

great many offerings at the feet of the wealthy shooter to tempt him. Some are bizarre and impractical, others work well but did not catch on and some survive to tempt the user of old guns. Here we shall investigate the bizarre and consider their use in the field, though it is a sad fact that most examples of the guns we encounter in this category are now museum pieces or form part of a collection. Many are unlikely to be used again for their original purpose.

While it is true that multi-barrelled guns are still made and have a long and distinguished history (primarily as shotgun/rifle combinations), the multi-barrelled shotgun, as experimented with by 19th century gunmakers, never enjoyed any high degree of success in Great Britain.

Green of Cheltenham, a true gunmaker with a number of important patents to his name, patented a three-barrel gun in 1902 with two conventional side-by-side barrels and a third on the top. Green built this gun himself. It was also built by Lancaster and Westley Richards. John Dickson of Edinburgh had also produced a three-barrel gun (in 1882), in which the three barrels were side-by-side-by-side, making it rather ungainly in appearance. I handled a pair of these in 20-bore at Bonham's in London in 2005. At only 6lbs 7oz, they compare with a 12-bore double game gun for weight.

However, three-barrel guns never did replace the double as the shooter's favorite. Christopher Austyn in

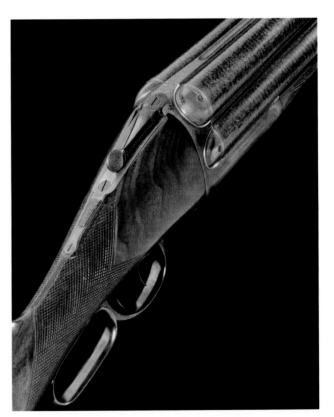

This Charles Lancaster 12-bore 4-barrel non-ejector is complex but works quite well. The triggers fire the top barrels and the under-lever cocks the tumblers and rotates them to fire the bottom barrels. The Damascus barrels are 28" and choked ¼ and ½. Weight is a hefty 10lb 14oz.

'*Modern Sporting Guns*' *(1994)* states that Dickson's are believed to have made only 27 three-barrel guns. Boss also made them as an experiment but their records show that only two made it as production models, both were 16-bores with the three barrels in line side-by-side-by-side.

Multi-barrel guns were clearly the objects of experimentation in the closing years of the 19th century and the beginning of the 20th. Gunmakers were angling to see what the public would buy and would provide anything that proved popular. Novel features and obscure patents proliferated and were offered to see if shooting men would 'bite'. Charles Lancaster patented several multi-barrel designs. Purdey only ever made one; a four-barrel 20-bore of beautiful quality (based on the Lancaster 1882 patent) which they built in 1883. The gun is very complex and looks rather ungainly in photographs, but it actually feels lively and well proportioned 'in the flesh'.

These exotic and obsolete designs are probably only of academic interest to most of us, but they do show some interesting 'blind alleys' down which gunmakers ventured in the search for the perfect balance between beauty, handling quality and rapidity of fire.

There were other attempts to increase firepower; some became phenomenally successful, like Browning's 1905 recoil-operated self-loading shotgun and the Winchester 1897 model pump-action, both of which were made in huge numbers until quite recently.

This Purdey four-barrel 20-bore built in 1883 and sold in 1886 to a French general, is the only multi-barrel gun on record at South Audley Street, it uses a Lancaster patent. I found it well balanced and pleasant.

Others were less successful and disappeared shortly after their launch. Of these, notable examples that one still occasionally finds at auction include the Spencer slide-action repeater of 1882 – very like a conventional pump-action in appearance apart from the secondary reversed trigger that cocks the hammer. These will be found

The Spencer Repeating Shotgun.

as imports to the UK with Damascus barrel and the name of Charles Lancaster on the action. Lancaster finished 70 of the guns when they arrived 'in the white' and distributed them in the UK. It is known that Rigby also imported the Spencer. Greener, in *Modern Shot Guns* referred to public trials given by renowned professional Shot, 'Doc' Carver, of the Spencer gun.

The trials were not conclusive, the guns jamming several times – Carver blaming faulty ammunition, a claim that Greener scorned. Having tested the Spencer (both Lancaster version and Spencer version) with modern plastic cased ammunition and finding it effortless and faultless with most shells I tried, (though occasionally failing to eject one brand of 70mm cartridge while shooting skeet), I am inclined to concur with Carver. Perhaps the design was a little too ahead of the ammunition available – paper cases are prone to more swelling and damage in the cycling of the action.

G.T. Teasdale-Buckell, editor of *Land & Water*, was sceptical about the emerging repeating shotgun designs. After trialing the Boss three-barrel single trigger gun at London Sporting Park, he commented 'We do not think with the late Lord Suffolk that repeaters will ever come in, although automatic loading guns may, some day, because of their absence of felt recoil, but at present the three barrel single trigger is the greatest advance that has been made to fowling pieces'. Writing in 1900, he was of course right about the degree of success the semi-auto would later achieve but his enthusiasm for the three-barrel gun with a single trigger was not shared by enough of his contemporaries to make it a commercial success.

A rival to the Spencer, the Winchester 1887 lever-action in 12-bore. The deep drop at comb requires a 'head-up' shooting style.

Winchester offered an under-lever operated repeating shotgun in 1887 – an action that will be familiar to cowboy film buffs. It was based on the famous rifle action manufactured by the firm. Even Greener conceded '...*the gun is neater in appearance than any repeating shot gun yet introduced and as a repeater does its work fairly well*'. I have shot the Winchester and found it less effective as a shotgun than the Spencer, which I quickly grew to like. The Winchester was improved in 1901 and nitroproofed, finding a degree of favour in 10-bore as a rugged wildfowling gun. The 12-bore guns I have tested have very little felt recoil. Earlier still, in 1860, Colt had adapted his famous revolver action for shotgun use and it did get some positive reviews in the British press but never became popular, largely due to safety considerations.

I see examples of the 'obsolete' shotguns described above quite regularly when I visit London auction houses, often in shootable condition. I must admit to a certain

admiration for the ingenuity of the makers of the time. They led the way for the development of the Browning 'long-recoil automatic, manufactured by Fabrique Nationale of Belgium, the first truly successful semi-automatic shotgun. They have an ugly charm of their own and as a revolutionary design, form part of shooting history. Brownings, Winchesters and Spencers currently sell for relatively little and as curiosities, as well as being very usable 'rough' guns, are worthy of consideration by any collector. In fact, all the transitional repeating shotguns offer an excellent opportunity to build an interesting working collection for a fraction of the cost of hand-built British guns.

All the guns described above can be had for a few hundred pounds in good condition, though they are harder to find in Britain than in the United States. I cannot see this continuing to be the case for very long. Perhaps we are about to see an increase in appreciation for these enchanting devices in the manner of the hammer gun revival of the last few years.

Engraving

I will not dwell on engraving styles in detail, as there are entire books dedicated to just this subject. Engraving can affect the cost of a new gun hugely and residual costs will be passed on to the second-hand market. Engraving has featured on sporting arms for centuries and was initially a simple embellishment applied before the gun was released to the buyer; the engraver just seen as a tradesman. Engraving

This Smith double flintlock of excellent quality displays the simple engraving typical of the era.

The Browning 'long-recoil' semi automatic shotgun.

gradually became more important to the average buyer of a sporting gun, but only relatively recently have the names of the engravers become of interest.

The development of the British sporting gun was pushed primarily by practical demands. Sportsmen used their guns hard and wanted faultless performance under demanding conditions. Every aspect of performance was tested to the extreme and weaknesses were quickly exposed; the proving ground of the pheasant drive or the salt marsh quickly delivered a verdict on new designs. Those that did not do best service fell by the wayside.

The British sporting gun is beautiful because it is the perfect development of metal and wood taking form for a specific task. The proportion, the lock work, the barrels and the stock are what they are because they do the job best. Engraving was seen as mere embellishment by some and of more interest to collectors than it was to many practical shooting men of yesteryear. However, the vintage guns we buy and use today each carry their own distinctive legacy from the engravers of old. It imbues them with much of their beauty and character.

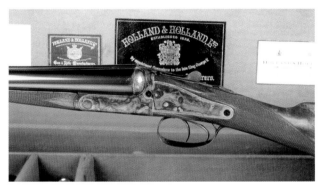

These Holland & Holland back-action guns from around 1930 show color case hardening (top) and the bare metal after the case hardening has worn away with use (bottom). Even the simple border engraving employed here 'lifts' the austere appearance of these lower grade guns designed for hard use overseas.

Engraving has a long history in the decoration of guns but has always been secondary to the practicality of the weapon in Britain. British makers seldom went in for heavy carving of the woodwork and the inlaying of gold and silver; continental makers were far more prone to this. Likewise, many British guns of the very finest quality are bereft of flashy engraving. Instead they are subtle and restrained.

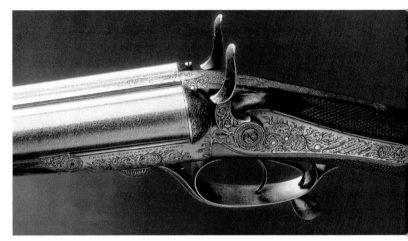

This Charles Lancaster shows the larger scrolls popular on early breech-loaders. Note the beautifully engraved nose-less hammers, typical of early Lancasters.

Bare metal shines in the sun. Engraving diverts the reflected light and diminishes the shine – a useful characteristic if you do not want to be spotted by keen eyes. Engraving does therefore have a practical application. When color case hardening has worn away from the action, the stark appearance of a bare, metallic surface is lessened by the presence of engraving. Perhaps this was influential in its development, perhaps not.

British sporting guns in the 1830s typically had a few large scrolls and a border pattern around the lock plates and hammers. During the middle years of the century the scrolls engraved on the lock plates began to get smaller; tighter and more profuse. The Purdey 'house style' of bouquets of roses and fine scrolls (termed '*standard fine*' at Audley House) was developed in the 1870s by James Lucas, according to Donald Dallas, who researched the subject while working on his definitive history of the firm. 'Lucas rose and scroll' is very fine indeed and, as Purdey was a leading gunmaker, lesser-ranked gunmakers copied his style and rose and scroll became fashionable. It is now recognizable as very 'English'. Jack Sumner, for Boss, developed his own interpretation of heavy-coverage rose and scroll while Holland and Holland took a bolder turn with their 'Royal' engraving of deep-cut stylized foliage and scrolls.

Fashion continued to influence the engravers and while there was always a preponderance of the traditional styles, more unusual patterns will be encountered on guns from the closing decade of the 19th century and the first three of the 20th. The Arts & Crafts movement influenced gun engraving, as it did furnishings and architecture. Celtic strap-work was popular with some, oak leaves, vine leaves, flowers or thistles with others and game scenes moved from being naïve and stylized, formed by line cutting, to the modern realistic scenes pioneered by the Italians and made from thousands of tiny dots.

A significant number of customers steadfastly refused to pay for engraving. They ordered 'best guns from the best makers but ordered them plain or with simple border engraving only. The untutored eye may be deceived into thinking the profusion of engraving on an English gun is a reliable indication of the quality of the piece. This

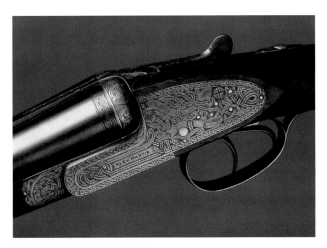

This 1935 Alex Martin sidelock shows Celtic style straps worked into a pattern as an alternative to scrolls, leaves or flowers.

is true only up to a point; the *quality* of the engraving is a better indication, as poor guns were not usually covered in expensive engraving by the best engravers. Teasdale Buckell writing in *Land & Water* in 1889 warned readers about this when he wrote *'As ivy is the bad mason's best friend, so on second rate guns does gaudy engraving conceal mediocre work'*. One regularly encounters guns of the very best quality totally bereft of engraving or featuring plain

locks but for a beaded or lined border. Many Holland & Holland 'Royals' I have seen are of this type and I have also seen them with blacked locks and actions. Westley Richards guns one encounters are often of fine quality but lack a profusion of engraving. Their 'Gold Name' boxlock is just that – a plain action with 'Westley Richards' inlaid in gold lettering.

From the 1870s onwards, gunmakers produced special guns to order or made exhibition pieces that were heavily and specially engraved. The 'St George' guns of W.W. Greener are a good case in point. They featured chiselled relief scenes of St George slaying the dragon. Similarly exquisite pieces were produced by other makers, but they were not the norm. While Indian princes ordered highly decorated guns of all descriptions, most sportsmen bought most of their guns to shoot with and practicality was uppermost. For most, engraving was still secondary.

Extra engraving was always an option and became increasingly popular with wealthy clients. It could, and can still, increase the final cost of a gun significantly. Highly finished guns for 'live pigeon' shooting are a good example of this. The social aspect of shooting among the wealthy classes allowed plenty of time for the examination and appreciation of other men's weapons and those so inclined were prone to demand a gun that would demonstrate their wealth and taste in a 'louder' manner than had been the norm among game shooters. Engraving played a key part in this and became ever more specialised until the present day, where a gun commissioned by a top London maker may cost as much to engrave and inlay as to build. The most famous modern engravers now sign their work and are booked months in advance by all the top gunmaking firms.

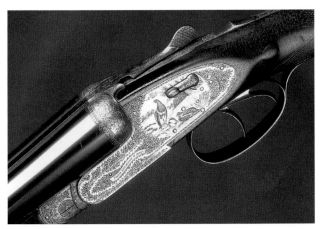

This 1924 Dickson sidelock shows tight scrolls and a naïve vignette of partridges.

Engraving is a matter of personal taste. When you see a gun, you either like the engraving or you do not. As well as the engraving of actions and lock plates, fences can be carved into many shapes, many indicative of the age of the gun and the fashion at the time. One could write a whole book on fences, but this is not the place to explore them in detail.

With experience it is possible to get a reasonably reliable 'feel' for the age and maker of a gun by the engraving style and the carving of the fences. Fleurs-de-Lys fences are common on 1880s Purdey sidelocks, Woodward often employed arcaded fences and W&C Scott put carved oak leaves and acorn designs on some of their 'Premier' grade guns. Oak leaf fences are also common on 1930s Purdeys. Acanthus leaves were popular at the turn of the century and Grant guns are often recognizable at a distance from their 'fluted' fences. However, there really is infinite variety and guns will be encountered with every conceivable theme featured in the engraving. This is part of the delight in visiting auction rooms and gun shops; you never know what you will encounter.

Gunstock issues

When a vintage British shotgun was originally made in the factory or workshop, it was fitted with wood according to the taste of the gunmaker. Although some gunmakers would select figured wood and some customers would specify it, in many cases it seems that the figure in a gun-stock was largely a matter of luck. In fact, many of Purdey's best guns have very plain wood, possibly selected for the straightness of the grain and the greater strength inherent in such wood, aesthetics taking second place in the mind of the buyer at the time.

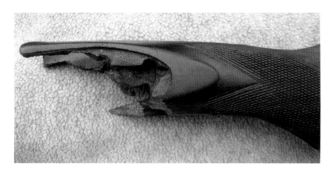

Typical damage to the delicate under-horns of a sidelock stock. Such damage requires a total re-stocking job.

However, certain discerning buyers have clearly always appreciated the beauty that a well-figured piece of wood can add to the finished article and over the years a number of woods have been used for shotgun stocks. Birds-eye maple was once used, though mostly for exhibition guns, and it is very beautiful but not as hard-wearing as walnut, or as easy to work (stockers hated it) and with age it becomes brittle. Beech is used for rifle stocks and may be found on some cheap shotguns but walnut is by far the most popular choice for the stocking of shotguns.

A friend of mine owns a William Smith converted flint gun stocked in fiddle-back (or tiger-stripe) maple, which was also favoured by some makers in the 1820s. Two more examples were sold by Christie's in 2000, from the W. Keith Neal collection. However, walnut was always the favorite of the gun trade: it has all the desirable qualities in abundance; it is easy to cut, holds checkering well, has a good strength-to-weight ratio and is naturally beautiful.

This 1823 Purdey percussion muzzle-loader shows the beauty of birds-eye maple as a gun stock material. I have never seen it used on a breechloader, though I am told the Royal Gun Room at Sandringham has a pair of 16-bore Purdey hammer guns so stocked.

The most popular wood traditionally used for English gun-stocks was French walnut but this is practically unavailable today and has largely been replaced by walnut from Turkey. French walnut is notable for the vanilla tones in the wood and the striking black contrast of the figure. Turkish walnut is darker and typically shot through with black stripes.

The most desirable wood, that with the contrasting figure, comes from the base of the tree, where trunk and roots join. The selection of wood is full of financial risk – it is often bought in the rough state and the amount of quality blanks obtainable from each trunk is not guaranteed.

Any flaws (called 'shakes') in the timber may not become apparent until cut into blanks at the timber mill or even later, in the hands of the stocker. At this point a valuable blank can become scrap.

It is likely that the collector of vintage guns may consider buying one in need of re-stocking or may be unfortunate enough to have a stock break in the field. Re-stocking is carried out exactly as if the gun were new and being stocked for the first time. The same options are available to the owner. A new stock can be made to the measurements of the owner, to his taste regarding grip, figure and butt plate – and of course, wood.

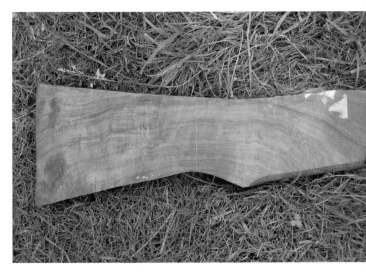

A Turkish walnut stock blank.

A Turkish walnut root freshly grubbed out of the ground and ready for the timber mill. The best, most highly-figured wood comes from this part of the tree.

The wood will typically come from a walnut root grubbed out of the ground and cut into 'blanks'. The blanks are purchased and kept for years to season until the moisture content and weight has stabilized. When ready for use, the blanks can be selected for use and marked for shaping by the stocker. The decision about how to cut a blank is crucial and a skilled job in its own right. The hand of the stock, where the gun will flex upon firing, needs to have a straight grain for strength and the wider butt-stock needs to contain the figure that is so aesthetically pleasing. If the re-stocking is to be really perfect, the forend wood should also be replaced to match the new butt-stock. This apparently minor job is actually quite difficult and costly.

The inletting of the action into a piece of wood is a precise and highly skilled task and the fine tolerances that top craftsmen are able to work to are unbelievable to those less skilled in the use of tools. In the case of the best work the line between wood and metal is as fine as if drawn on with a sharp pencil.

Restocking an old gun will detract from its appeal

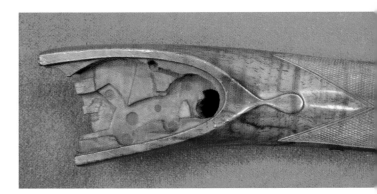

A walnut stock cut to receive the lock-work of a Purdey sidelock.

to many collectors but to the user of old guns, re-stocking a fine old action can be an excellent means of getting a bespoke gun at budget prices. The work is not cheap – $1,000 for the wood plus $4,000 for the work is what a best quality job will cost on a sidelock. Boxlocks cost less and bar-in-wood hammer guns more. However, when finished, the gun will fit you perfectly and be exactly as you decide you want it.

To make sure of perfect dimensions, it is imperative to have a proper fitting with a shooting instructor at a shooting ground. Measurements are not simply physical – eye dominance issues also affect the amount of cast required and firing at moving clays as well as at the pattern plate will be part of the evaluation process. It may cost you $200 but it will be money well spent.

Stock shapes and grips

Stock shapes vary more than most people realize: in vintage guns as in modern ones. Height and thickness of comb, cast and drop, angle and contour of butt and type of butt plate will all reflect something of the age and quality of a gun or the peculiarities of its original or subsequent owners. However, there are certain features that are common enough and recognizable enough to be listed and illustrated and referred to as 'types'. These are as follow:

The straight-hand stock

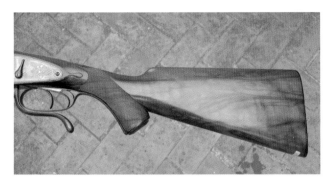

The pistol-grip stock

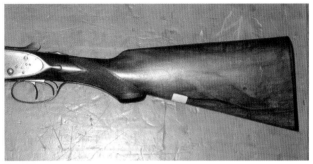

The 'Prince of Wales' stock

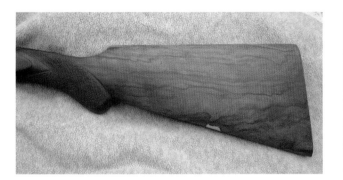

The semi-pistol-grip stock

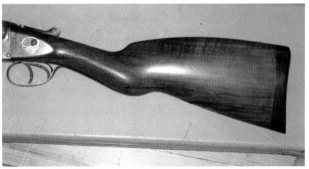

The Greener 'Rational' stock

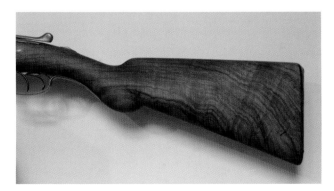

The swan-necked stock

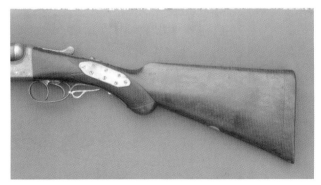

A stock with a cracked and repaired hand and wooden extension

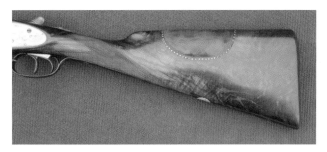

A stock with a leather cheek-pad

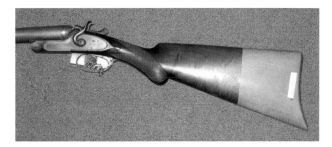

A semi-pistol grip stock with ill-advised beech extension

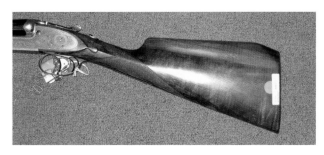

A stock with a high Monte Carlo-style comb extension

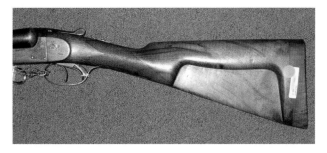

A lightweight gun stock with reduced dimensions

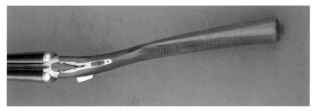

A cross-over stock

What to collect and use

What is the best gun? Everybody has the right answer to the question.

Greener had the right answer in 1909, Major Burrard had the right answer in 1932, Mike George had the right answer in 1998 and the chap you speak to in your local gun shop will have the right answer now – uncannily, he will even have just the thing in stock.

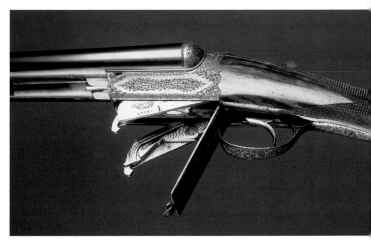

Not all 'best' guns are London sidelocks. This Westley Richards with hand-detachable locks, patented by Taylor and Deeley in 1897, is as beautifully crafted as anything the London trade offered.

The funny thing is – none of the above agree! We will explore what they said and why they reached their conclusions later on. But I would suggest that the best gun is the one that suits you best. You need to know your own mind. It will be useful at this point to define the word 'best' because it has a peculiar usage in gunmaking. 'Best' guns are those of the absolute top handmade quality, made without thought to cost. When exploring the best gun for a given purpose I shall use *best* rather than 'best'.

1858	1859	1861	1862	1863	1864	1865	1866	
Westley Richards Doll's Head Top Lever	Jones Screw Grip Rotary Under-Lever	Centre Fire guns	Westley Richards Doll's Head	Stanton Rebounding Lock

Purdey Bolt

Thumbhole Lever | Powell Lift-up Top-lever

Westley Richards Sliding Top-lever | Scott Spindle | Pape's Choke | |
| Gun developments | Breechloaders become reliable | | | | | | | |
| Powder developments | | | | | | | Schultze powder Smokeless powders | |

Burrard, an artillery officer who shot game in many a far-flung outpost of the British Empire, is one of my absolutely favorite writers. Macdonald Hastings, rather uncharitably, described him as a 'fussy old gentleman who did not know much about what was worth reading' but Burrard, though certainly not infallible and guilty of indulging in feuds with at least two of his contemporaries (Hugh Pollard and Robert Churchill), never fudged an issue. His analysis is forensic and absolutely informed. You may not like his mode of qualitative evaluation but you cannot fault his correctness, attention to detail or precision.

Burrard's detractors are generally those who have no patience with his comparative, scientific style (an absolutely acceptable opinion in the sphere of personal gun choice, which has an artistic as well as a scientific point of view to consider; though Burrard does refer to aesthetics, if rather coyly, in his evaluation). His trilogy *The Modern Shotgun* is essential reading for the serious student of shotguns and his little volume *In the Gun Room* is extremely clear and direct in answering many common shooting questions and exploding common myths.

Technical issues

Before this section commences, I believe it politic to warn the reader that it cannot go into enough detail or cover the entire range of the hundreds of patents for improvements in gun design that will be encountered at auction viewings or through the examination of guns from any of the many sources available to the potential purchaser or collector.

For the technically inquisitive there are excellent reference books by Crudginton and Baker, Geoffrey Boothroyd, 'Gough Thomas' and others. It is to these tomes, which are listed in the bibliography, that I direct the inquisitive collector.

My intention here is to uncover some of the more commonly encountered and influential designs and features, which the seeker of vintage guns is likely to need to recognise and evaluate when choosing his guns and deciding on their usability or otherwise for his intended purpose. My coverage is intentionally simple and pragmatic, for we are here to consider what matters and how it works rather than become over-involved in scientific, historical and technical minutiae. These matters are expertly described by authors better equipped than this one to provide the forensic examination of the technical and historical data required by the true scholar. Crudgington and Baker's exhaustive history *The British Shotgun*, in two volumes (with a third on the way) will be of immense value to those wishing to explore this fascinating subject further.

Readers should find in the following pages information of use in dating guns through certain features they may possess, to recognise the merits and limitations of the designs encountered and evaluate how they may affect the ability of a particular gun to perform a particular function in today's shooting field. It is aimed at the potential buyer and user of the guns described.

I have devised 'technical timelines' to help the reader quickly narrow down the date of manufacture of his gun through the visible reference points it provides the examiner. This should make it entirely possible for the layman to describe and date any commonly encountered gun of the 1859–1900 period to within a few years.

As a young man, when examining a gun, I would ask myself a number of questions but found I was unable to answer enough of them to satisfy my curiosity. They were:

Is the gun safe to use?

How old is it?

What was it made for?

What are the levers and lock-plates telling me about it?

Is it a good quality piece?

What is it worth?

Who was the man who made it? (More likely 'men'.)

I hope to equip the reader with the rudiments of evaluative knowledge needed to answer most, if not all, of these questions.

1867	1871	1872	1873	1874	1875	1876	1878	1879	1880	1882	1884	1889	1894	1897	1909
Rebounding Lock perfected	Murcott 'Mousetrap'	Anson Push-Rod forend	Greener 'Treble Grip Fast'	Needham Hammerless gun	Anson & Deeley Boxlock	Woodward 'Automatic'	Scott Back-action	MacNaughton gun	Greener 'Facile Princeps'	Webley Screw Grip	Bentley 'Semi Hammerless'	Southgate Ejector	H&H 'Royal' perfected	Westley Richards Detachable Lock	Boss over / under
			Deeley & Edge forend	Greener 'Treble Wedge Fast'	Popularising of choke boring		Hackett Snap-forend	Rigby/Bissel Rising Bite	Dickson 'Round Action' Beesley / Purdey Sidelock		Holland & Holland 'Royal'		**Boss Single Trigger**		

	Hammerless guns				**Reliable Hammerless guns**						**Decline in Hammer gun production**				
									E.C.Powder				Sporting Ballistite 'Cordite'		

GUN ACTIONS
1. Descriptive

Vintage guns of interest to the modern shooter can be roughly divided into two categories:

    1. Hammer guns

    2. Hammerless guns

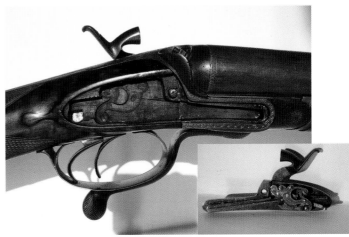

These pictures illustrate the internal components of the bar-action hammer lock (inset) and the recesses cut into the wood and the bar of the action to accommodate the internal mechanism. The gun has re-bounding locks and is by J. Thompson and dates from the late 1870s.

Left: Purdey pigeon hammergun. Right: Purdey hammerless sidelock. Both guns are twelve bores and use bar-action sidelocks.

The terms themselves are inaccurate because all guns have hammers really, either they are on the inside of the lock (and called 'tumblers') or they are positioned on the outside of it and called 'hammers'.

    Hammer gun locks fall into two main types: 'bar-action' locks (in which the mainspring is housed in a recess cut into the metal 'bar' of the action) and back-action locks (in which the mainspring is housed behind the action body on a lock-plate which is inlet into the stock just behind the hammers). Be aware though that some back-action locks can look rather like bar actions, even though the mainspring is not located in the bar. The position of the pins, seen on the outside of the lock plate, is the giveaway.

Types of hammer gun

Two main distinctions can be made regarding hammer guns, as stated above:

    • **Bar-Action Hammer Guns**, in which the mainspring is housed in a portion of the lock plate forward and below the hammer. The bar of the action is cut away to allow the spring to be recessed into the space provided.

    • **Back-Action Hammer Guns**, in which the mainspring is housed in a portion of the lock plate behind the hammer. Wood is cut away from the stock to receive the lock, behind the action body.

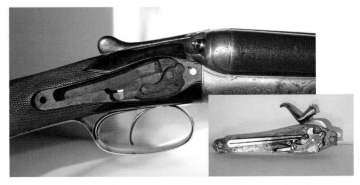

Here can be seen the internal components of a back-action hammer lock (inset) and the recess in the stock cut away to receive the mechanism. This is an 1885 Holland & Holland with rebounding locks.

Above left: a conventional bar-action hammer gun circa 1880, by Gallyon. On the right is a late 1870s back-action hammer gun by Purdey. Note the isolated lock plates.

Above: a direct comparison of bar action and back action hammer locks. On the left is a Robert Adams lock from a bar-in-wood hammer gun circa 1868. On the right is an Adams & Co circa 1885. Both have rebounding locks.

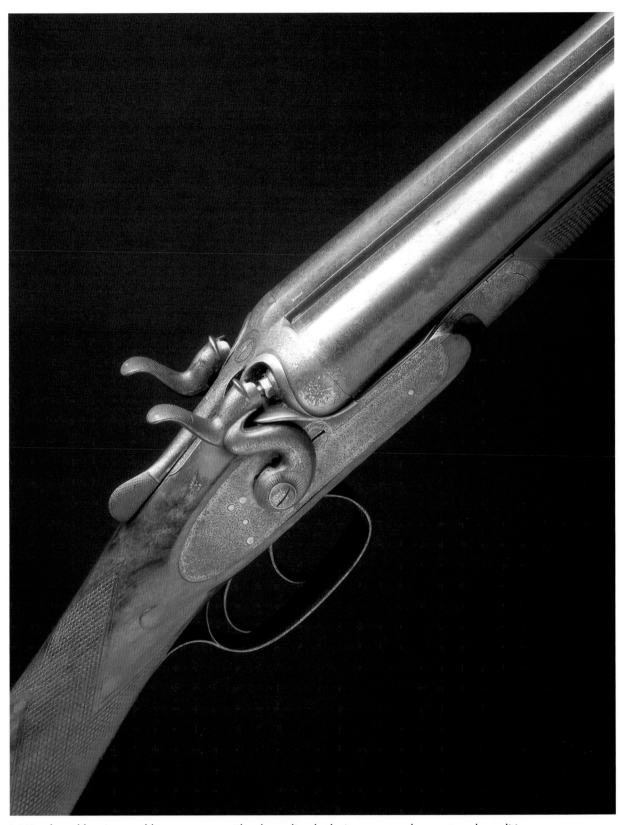

J. Woodward bar-in-wood hammer gun, with rebounding locks in very good, unrestored condition.

This Stephen Grant hammer gun with rebounding locks (right) shows the hammer in 'fired' position, resting just above the striker. The Lang bar-action hammer gun with non-rebounding locks (left), also in the 'fired' position, shows the hammer still in hard contact with the striker.

Two further distinctions are useful to make:

• **Rebounding Locks,** from 1866-1867 patents by Stanton and others introduced the useful improvement in hammer guns of the hammer springing back to rest in a position immediately above the striker after firing, rather than coming to rest upon it. The gun could then be opened without further manipulation of the hammer.

• **Non-Rebounding Locks.** Before 1867 (and in some later guns, as the rebounding lock was not universally adopted overnight) the hammers come to rest on the strikers and must be pulled to 'half-cock' before opening the gun (necessary to relieve the hammer-pressure on the strikers, which will be embedded in the cap of the fired cartridge).

The Bar-in-Wood (or Wood-Bar) Action

A further feature of some hammer guns is worthy of comment, for it is visually notable. This is the bar-in-wood action. This is, in fact, a bar-action lock but one in which the wood of the stock is extended to encapsulate some or all the metal of the bar. It is an attractive feature and commonly encountered in Westley Richards guns of the 1860s and Purdey guns as late as the 1880s, though they are also encountered by many other makers of the period such as Thomas Horsley and William Powell.

Above: Hammers shown at 'half-cock' on a non-rebounding lock hammer gun by Thomas Sylven, sold in 1868, a year after Stanton's patent for rebounding locks. Note that the 'half cock' position of the hammers provides greater distance between hammer nose and striker than is the case in a rebound lock.

Coil Spring Locks
– not such a new idea

Most English guns made before the 1960s had locks which operated on leaf springs. These were either mounted on the lock-plates of a sidelock like the Holland & Holland 'Royal', housed within the action, as in the Anson & Deeley or mounted on the trigger plate, like the Dickson 'round action'.

However, 19th century gun makers' began incorporating coil spring locks even in the early days of hammerless shotguns. Leaf springs have two main failings: They are quite time-consuming and expensive to make properly. They also suffer from the fact that when they break, they do so suddenly, resulting in the total disabling of the mechanism.

Profits improve by lowering the cost of production and shooters are never best pleased when their day is brought to an abrupt end by mechanical failure. Though merely inconvenient when bird shooting, if your mainspring breaks while taking aim at a charging lion, it could prove terminal. Another consideration for colony-bound sportsmen was that replacing a broken mainspring requires some basic gun-smithing tools and skills that were not always available on safari or in remote hill stations.

Coil springs still work when broken. In fact, they can be broken in more than one place and still function. This makes spring breakage a cue for replacement as soon as prac-tical but does not stop you from shooting. They are also faster and easier to make, adjust or replace.

The downside of coil springs is that they do not produce the rapid 'snap' action or the instant power of a good leaf spring. Many gunsmiths believe that coil-spring locks do not offer the same quality of trigger-pull and detonation that leaf springs do. Nevertheless, coil springs appeared in a few notable patents from the 1870s.

The spiral-spring Woodward of 1876

Thomas Woodward patented a hammerless push-forward under-lever snap action gun (patent 651) in 1876. The first Thomas Woodward, trading from a Birmingham premises, was the father of a second Thomas Woodward, who also traded in Birmingham before moving to London to manage the new Holland & Holland factory in 1893.

The 1876 Woodward patent was among the first workable hammerless actions, coming one year after the Anson & Deeley and only five years after Murcott's famous 'Mousetrap' showed that hammerless guns were things with a future. Woodward's patent was adopted by a number of makers, like Adams & Co of London and known as 'The Acme'. J. Beattie & Co of London sold it in 1882 in four different qualities, from 'Best' at $68 to 'Community' at $51.

Spiral springs mounted on the top strap of the Wood-ward 1882 patent.

Spiral springs like these are cheaper to make than leaf springs.

The 1882 patent of T&T Woodward

The Woodwards took one step further with their 1882 (2344) patent, enabling their spiral-spring locks to work with a top-lever. As well as having top-strap mounted coil mainsprings, it had twin coil springs fitted around rods, which functioned as sears, activated by the trigger blades. However, the big coil mainsprings required a lot of wood to be thinned either side of it. This weakened a vulnerable jointing area and the guns were prone to damage or to becoming loose.

Crudgington & Baker refer only to patent descriptions of 2344 but I own a working example, bought in a provincial auction and am pleased to confirm that the design was actually manufactured by T. Woodward. If this were a postage stamp, it would command a hefty price. It is interesting that scarcity does not affect gun prices in the way it does so many other objects of collection: my unusual Woodward cost me just £65.

Externally, the gun resembles a Dickson 'round action' with its solid, rounded bar and straight-backed action. It has 30" Damascus barrels and a top-lever opening mechanism with Purdey under-bolts and a 'doll's head' rib extension.

Woodward was not the only gun maker building spiral-spring doubles at the time. Stephen Grant guns dating from the 1880s regularly appear with trigger plate locks fitted with spiral main springs. The Horatio Phillips patent is especially reminiscent of the Dickson externally.

Robertson's coil-spring ejector for Boss

There is a degree of logic in having coil-spring ejectors that work in the same direction as the ejected shell case travels. The Boss ejector does exactly that. When tripped, it pushes the extractor forcefully upwards, as the spring expands, and ejects the case. This ejector is easily regulated and very reliable. The repair and replacement of these springs is routine and there is no danger of sudden failure, unlike those in the (justifiably) much-vaunted 'Southgate' system, which uses leaf-springs.

Holland & Holland's self-opening mechanism

Holland & Holland opted for a separate coil-spring, located under the forend and entirely independent of the locks to operate their 1922 self-opener. This performs the task of forcing the barrels open when the top-lever is operated and is also very reliable, easy to adjust, maintain and replace.

Whilst coil-spring operated ejectors and self-opening mechanisms were adopted by Boss and Holland & Holland, both still favoured leaf springs to operate their locks. However, while Webley made large numbers of back-action sidelocks with coil springs, commonly found on guns bearing the 'Army & Navy' name, leaf springs were preferred for 'best' guns.

20th century 'Budget' guns

The coil spring lock idea was resurrected, with patents between 1904 and 1917, by Charles Rylands, made until the 1930s as the 'Target' guns of Ward & Sons, with their distinctive conical fences. These low-priced boxlocks look and function much like the A&D but there is only one externally visible pin on the action. Cogswell & Harrison's 1922 'Moore-grey' also featured coil-springs in their back-action side-locks but quality was poor and the model short-lived.

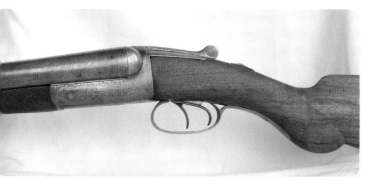

The spiral spring Woodward. Outwardly it is reminiscent of the round-action Dickson.

The 1876 patent spiral-spring Woodward on a snap under-lever gun by Adams.

Two Birmingham-made boxlocks of very different grade and style.

Top: a Westley Richards (Anson & Deeley action) made in 1897 of very high quality and finish. This would have been a very expensive gun, made to the highest standards. It has typical Westley Richards bolted top-lever and scalloped 'fancy back' action.

Below: a plain but sound Greener non-ejector of 1892, built on the 'Facile Princeps' (Greener's own 1880 patent) action. Despite its lack of engraving and finish, the wood-to-metal fit is very good and the gun is tight and serviceable despite a century of use. The non-automatic side safety is a Greener feature.

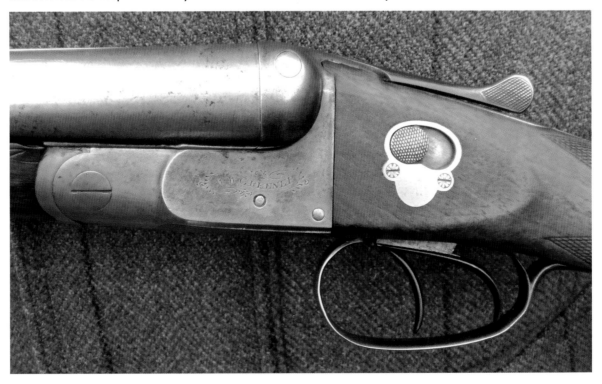

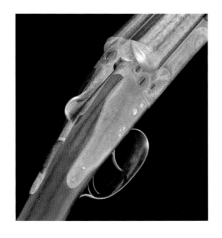

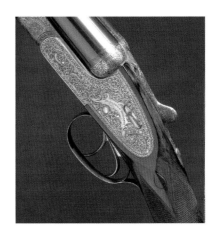

Types of Hammerless Gun

Comparison of common hammerless actions encountered: (left to right) a back-action sidelock (Holland & Holland), a boxlock (Westley Richards), a bar-action sidelock (Holland & Holland Royal).

Hammerless side-by-side guns fall into three basic types: boxlocks, sidelocks and trigger-plate locks (see above). By far the most common are boxlock and sidelock variants.

Boxlocks

Boxlocks are basically variants of the 1875 Anson & Deeley action in which the mainsprings are cocked by the fall of the barrels and sidelocks are basically hammer guns of bar-action or back-action form with the hammers on the inside of the lock, rather than the outside. Early hammerless sidelocks were cocked by the manipulation of the opening lever; later guns utilized the fall or rise of the barrels to cock the tumblers. Boxlocks are cocked by the fall of the barrels.

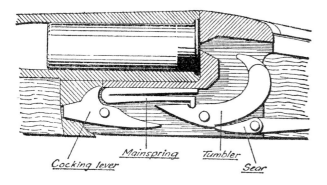

The Anson & Deeley boxlock of 1875. Robust, simple, and with only four moving parts: a true marvel of Victorian invention.

The best-known variations on the boxlock theme are those of Birmingham gunmaker and prolific writer W.W. Greener – known as the 'Facile Princeps' (meaning 'easily the Chief', in typically self-aggrandizing Greener style) and 'Unique' actions. As with almost all pre-WW2 Greener guns, they are invariably well made, even over-engineered, and function reliably and efficiently for years with minimal servicing.

Probably the finest development of the boxlock (I can hear Greener turning in his grave) is to be found in the Westley Richards 'drop lock', patented in 1897 by Deeley and Taylor. The locks are mounted on plates and inserted from the underside of the bar, via a hinged plate. This not only aids removal of the locks for security, cleaning or replacement in case of malfunction, it also removes the need to drill holes into the action bar to receive the tumbler peg, thus adding to the strength of the action. Burrard objected to the tumbler-peg hole as a point of significant weakness in the Anson & Deeley design, which he declared weaker than bar-action or back-action sidelocks when he evaluated their comparative merits, though in practice the strength of a well-made boxlock has long been established.

Sidelocks

Sidelocks vary in their design far more than boxlocks. Those that have stood the test of time and are the best known are the Boss, Purdey and Holland & Holland patterns. At the time of writing all three are still in production and all are well over 100 years old. However, many other sidelock

Above: a Purdey bar action hammerless lock (from a 1929 competition 'live pigeon' gun) and the recess in the stock and action bar made to accommodate the lock-work. Note the main spring in front of the tumbler and the hollowed-out action bar in which it is housed.

Below: a back action hammerless lock from a Williams & Powell game gun. The recess in the action bar houses only the cocking lever, as the spring is behind the tumbler and housed in a recess cut into the wood.

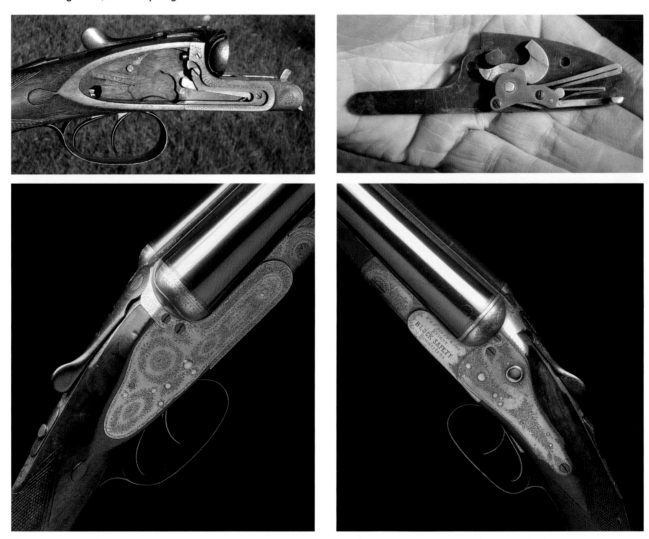

A direct comparison of two successful hammerless sidelocks from the transitional period before the industry accepted the stocked-to-the-fences bar action hammerless type as the action of choice for 'best' guns. On the left is a first model (1883 patent) Holland & Holland 'Royal' bar action sidelock. On the right is the Scott & Baker (1878 patent) back action sidelock made by W & C Scott.

designs were extremely successful, every major maker seeming to have his own patent or variation. The Webley & Rogers pattern, for example, is widely encountered in the guns of William Evans and Army & Navy CSL, and was made in large numbers.

The sidelock is, as mentioned earlier, essentially a hammer gun – but one in which the hammers are placed inside the lock rather than outside it. They are still basically working on the same principle: hammers fall, hit the strikers and detonate the cartridge. The lock plates of a sidelock are where all the lock-work is mounted, as is the case with hammer guns.

Sidelocks can be divided, as can hammer guns, into:

- **Back-action sidelocks**
- **Bar-action sidelocks**

This Edwinson Green over-and-under features back-action sidelocks. Note the mainspring behind the tumbler.

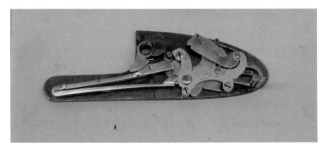

This Purdey bar-action sidelock shows the mainspring in front of the tumbler.

The same definitions apply regarding forward or rearward placement of the mainspring as previously outlined with regard to hammer guns. It is worthy of note that all 'sidelock' over-and-under guns use back-action

locks, though at the time of writing the Charles Hellis firm is developing a true bar-action, self-opening, over-and-under sidelock based on the Beesley/Purdey.

Trigger-plate locks

Less frequently employed for side-by-side guns is the trigger-plate action, which is widely used by modern gunmakers, including Beretta and Holland & Holland, for over-and-under guns. It is usually encountered in vintage side-by-sides patented by the Scottish makers John Dickson (1882) and James MacNaughton (1879). The Scottish tradition for trigger-plate action side-by-sides continues today with the guns of David McKay Brown of Bothwell, near Glasgow. These guns are well-balanced and the actions can be filed down to graceful proportions, while retaining the balance point around the hinge-pin. The tumblers are generally cocked by the fall of the barrels, though sometimes (early MacNaughton guns) by the movement of the opening top-lever.

Burrard wrote favorably about the Dickson in particular. He was especially impressed with the strength of the action and the fine trigger pulls. In summary he writes:

'The workmanship is faultless, and the design of the lock such that it is as good and efficient as the bar-action sidelock. I do not see that it is any better than a bar action sidelock but it is certainly as good, although I think the addition of an intercepting safety should be made [unlike all the 'best' sidelocks offered by top firms, the Dickson round-action lacks a device that blocks the fall of the tumbler when the safety is engaged – the Dickson safety is merely a trigger-lock]. *Its great merit, however lies in the strength of the bar; and on this account it is particularly well adapted for the building of very light guns.'*

The Dickson round action makes an interesting alternative to a conventional sidelock. Only around 2,000 have been made to date and they have a good reputation among gunmakers, who appreciate their qualities and among Scots and their descendents whose sense of national pride is gladdened by carrying a Scottish gun in the field and knowing it to be as good as anything to come out of a London or Birmingham workshop.

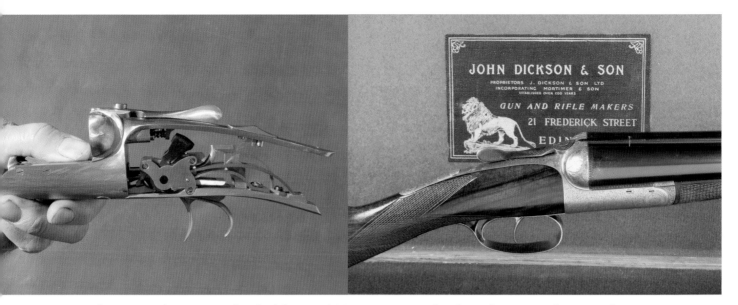

Above: A modern trigger-plate lock by David McKay Brown. Right: The Dickson 'Round Action' of 1882 a trigger-plate gun on which the McKay Brown is based.

Table of Detonating Mechanisms							
1867	1871	1874	1875	1876	1879	1880	1884
Stanton rebounding lock	Murcott Mouse-trap	Needham Hammerless gun	Anson & Deeley Boxlock	Woodward 'Automatic'	MacNaughton 'Edinburgh' gun	Greener Facile Princeps Beesley/Purdey self-opeing sidelock. Dickson Round Action (1882)	Holland & Holland 'Royal'

GUN ACTIONS
2. Evaluative

Evaluating the merits of different gun locks

Many writers have discussed the merits of sidelock, boxlock and other designs in the quest for a definitive answer to the question 'Which is best?' Similar arguments raged when hammerless guns were introduced to compete with hammer guns for the attention and money of the shooting public.

Sidelock or Boxlock?

W.W Greener argued vigorously that the boxlock was superior to all others (his penchant for boxlocks led him to produce a number of variants of his own, mentioned earlier, such as the 'Facile Princeps' (1880) and the 'Unique' (1889)

– this resulted in a legal dispute between Westley Richards (owners of the Anson & Deeley Patent) & Greener, which went to the House of Lords – ultimately won by Greener on a technicality). In *The Gun and its Development* (the 9th edition, of 1910) Greener writes:

'The Anson & Deeley type of lock gives quicker ignition than the ordinary lock, for the blow is much shorter and the mainspring stronger. The sidelock hammerless guns do not have this advantage. Some of them also are liable to miss-fire, especially the lower priced ones – for the tumblers and other lock mechanism, being placed so far from the joint piece of the breech action, require long bolts and levers to effect the working of the locks and leverage being lost by the distance from the fulcrum, the tendency is to make the mainspring very light, in order that the cocking of the gun may seem easy and not cause the barrels to drag too heavily when the gun is opened. The advantage they possess is the ease with which the locks may be removed and the lock work inspected. This is not

a matter of importance, since a well-made boxlock is so placed as to be efficiently protected from the intrusion of dust, dirt or wet, and will work well for years without attention'.

Henry Sharp in *Modern Sporting Gunnery* (1909) was of the same mind as Greener and he reminds us that really successful hammerless guns emerged as boxlocks before they did as sidelocks:

'The side lock hammerless gun is an off-shoot of the Anson & Deeley barrel cocking hammerless method. Its external appearance is doubtless pleasing but this arrangement has been purchased at the sacrifice of the perfect simplicity of the original Anson & Deeley'.

Greener (*shown below*) continues on the theme:

'The great advantages the [boxlock] principle possesses over those in which side locks are used should determine the sportsman in his choice, for, in addition to the disadvantages already mentioned, side-lock guns are found to be more liable to accidental discharge: the

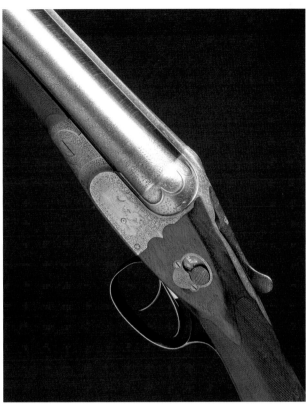

Although commonly used in guns in the medium and lower price brackets, due to the lower cost of production, boxlocks of extremely high quality such as this Greener 'G Grade' are not as uncommon as many think. Note the Greener side safety.

weaker lock mechanism more readily 'jarring off' and this, as recently proved, notwithstanding the safeguard of automatic intercepting locking bolts to scears, tumblers or triggers.'

Major Sir Gerald Burrard argued equally authoritatively for the superiority of the bar action sidelock. In his precise, analytical style, he outlined six factors to be considered when appraising the merits of the three hammerless lock types in general use.

1. STRENGTH. That is the degree in which each type leads to strength or weakness in the body of the action.

2. EASE IN COCKING. A very important point, as a lock which is difficult to cock means a gun which is stiff to open – a serious disadvantage when shooting fast.

3. EFFICIENCY. That is the efficiency of the lock in actually firing the cartridge, and so reducing the tendency to miss-fires.

4. QUICKNESS. By this is meant the length of time which elapses between the trigger being pressed and the gun being fired.

5. SAFETY.

6. DELICACY OF TRIGGER PULL. A smooth, crisp pull is a great advantage in shooting, while anything in the nature of a drag in the pull is fatal to good work.

Following his detailed evaluation of each lock in each of the six categories (which I shall not repeat here), Burrard concludes by conceding to the boxlock's equality in two areas:

'In two out of the six points, which combine to make a perfect lock – namely, 'Ease of Cocking' and 'Quickness' – all three types of lock [bar-action sidelock, back-action sidelock and boxlock] have been seen to be equally efficient.'

The sidelocks begin to edge ahead in the next two:

'In two more – 'Efficiency' and 'Safety' – the two types of sidelock are slightly superior to the boxlock, although when the latter are fitted with intercepting safeties they equal the sidelocks in this last respect.'

The Major finally damns the boxlock with the following observation:

'But in the matter of 'Strength' the boxlock is undoubtedly inferior to the sidelocks, while the back–action sidelock is the strongest of all.... And in the last, but important, question of trigger pull, the bar-action sidelock proved better than either of the others...'

Burrard concludes:

'It must, therefore, be admitted that the bar action sidelock is the best of all three, since it is as good, or better, than the other two types in five out of the six points, while in regard to strength it permits the use of an action which is quite sufficiently strong for normal use without the addition of a top extension...'

This is an assertion that Gough Thomas, writing in 1963, took issue with:

'he [Burrard] insisted that sidelocks are inherently stronger than boxlocks – an opinion that cannot be sustained on any fundamental grounds.'

Burrard also considered the bar-action sidelock more beautiful and graceful. Public opinion in the 75 years since he penned this opinion seems to support him. The convention for expensive guns has long been established and the bar-action sidelock is more usually encountered in 'best' side-by-side guns than any other type.

Greener himself made thousands of best quality sidelocks, despite the common perception of the typical Greener gun being a boxlock; this is partly fuelled by the famous 'St George Gun', of 'G-Grade' boxlock form, with distinctive chiselled engraving, found on the cover of his book *The Gun and its Development*. Greener was rightly renowned for his beautiful 'G-grade' boxlocks, built on his 'Unique' action by an elite team of Greener employees until the outbreak of The Great War. However, he was swimming against the tide and the sidelock was the design that gained greater acceptance as that most suited to build the finest guns upon.

The Quality Factor

Burrard and Greener were both careful to stress that their evaluations of superiority of one type over another were restricted to the consideration of like guns of high quality only. Another authority, Major Hugh Pollard, writing in 1923 commented:

'Sidelocks make the arm more expensive and in no way add to its efficiency, but they make it rather more delicate in balance, graceful in appearance, slenderer and more tapered at the action...'

Gough Thomas also noted that quality issues rather than chosen mechanism influenced the forging of reputations, both good and bad:

'it is almost certainly true that the majority of cracked

Regardless of the action type chosen, quality is the key factor and this mid-1880's, 6lb 8oz John Blanch back-action 12-bore has it in spades. Everything about the gun is 'best' quality despite it not being a bar-action sidelock.

actions [usually at the proof house or after abuse] are boxlocks, but this cannot be regarded as evidence of a fundamental weakness of boxlocks as a type, but the weakness of individual guns, as arising out of poor design, inadequate dimensions, inferior material, bad manufacture or, maybe, incorrect heat treatment.'

He goes on to state that this is hardly surprising if (as is the case) most boxlock guns were made in the lower grades. He also points out that W.J. Jeffery & Co. chose a boxlock action for their monstrous .600 double rifle, proved to withstand 14-ton service pressures; hardly an indication that eminent gunmakers believed the Anson & Deeley to be an inherently weak action.

However, it soon became apparent to gunmakers (after 1875) that it was cheaper to make a boxlock than a sidelock of comparative quality. Therefore, when a gun was made 'down to a price', pound-for-pound, the boxlock was likely to be the better gun. This goes some way to explain the high numbers of medium-grade guns of boxlock form available on the second-hand market.

The simplicity, durability and essentially service-free qualities of the boxlock made it popular for guns likely to see rough use or foreign travel. The back-action sidelock was also put to use for its strength and where heavy charges were to be used, such as in double rifles, but also when guns were made lightweight and strength in the bar became particularly important.

It is interesting to note that Holland & Holland settled on the bar-action sidelock for their *Royal* best grade gun (1884) and the back-action sidelock for their *Climax* (made by W&C Scott on the Scott & Baker patent of 1878) and *Dominion* (1934) guns, designed for hard use. However, the self-proclaimed *'builder of best guns only'* (Boss) made wide use of back action sidelocks for 'best shotguns'. An 1888 edition of 'Land & Water' quotes the then proprietor of Boss, Edward Paddison, advising that back-lock guns *'leave the action stronger'*. W.W. Greener agreed with this point and recommended back-locks for light weight guns. After Paddison's demise, in 1891, it is interesting to note that his less conservative successor at Boss, John Robertson, forged ahead with a bar action sidelock design, similar to the *Royal* in many respects, for which the firm is now rightly renowned.

As we have seen, the historical evaluation of gun actions and lock types is not conclusive in proving that one configuration is decidedly superior to another, though Burrard makes a very good attempt at working it out

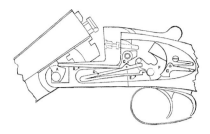

The famous Purdey self-opening sidelock. Invented by Frederick Beesley and sold to Purdey for $267 in November 1880. It has been used by Purdey for their side-by-side guns ever since. This ingenious design involves the tumblers (F) being cocked by the upper arm of the mainspring (D). When the gun is opened, the kickers (B) force the barrels open due to pressure from the mainspring (E). The gun literally opens itself once the top-lever is pressed. The mainspring is only put under tension when the gun is closed. It has a conveniently wide gape for easy reloading.

systematically. However, the market long ago decided that, in agreement with Burrard (though the convention was established long before he published *The Modern Shotgun* in 1930), the bar-action sidelock is the most desirable for high-grade guns.

If the bar-action hammerless sidelocks of Purdey, Boss and Holland & Holland pattern have become the most widely appreciated forms of the hammerless gun, it is worth noting that some very successful back-action designs enjoyed widespread acclaim and application. Notable is the Scott & Baker patent of 1878. This featured visible cocking rods on the action table, which cock the tumblers when the barrels drop. This action will be encountered on guns retailed by W & C Scott, Holland & Holland and Cogswell & Harrison, as well as others. As well as being robust and well proportioned, these guns are easily recognized from the distinctive shape of the lock-plates and their crystal windows through which the gold-washed tumblers can be seen when cocked. Whilst many of these guns were made plain, others, especially Scott-badged pigeon guns, are of exquisite quality and finish.

Bar-action sidelocks certainly do not have

Lever-cocking and barrel-cocking actions

Hammer guns were easy to operate because of their relative simplicity. The opening lever simply opened the gun, the bolting system fixed the breech end of the barrels to the action face, and the hammers were cocked manually once the gun was closed.

In the search for increased speed, the inventors of the day turned to the issue of automatic cocking of the hammers and a number of patents exist, such as the E. Hughes patent of 1878, in which the operation of the opening-lever also cocks the hammers in the same movement.

Once hammerless guns became the primary focus for those inventive minds, the issue of cocking the tumblers became more critical, since the shooter could not put his thumbs on them and pull them back to 'cock'. Automatic cocking became a necessity, not a novelty.

Two basic ideas were followed – Lever-cocking hammerless guns (Murcott made the first in 1871) and Barrel-cocking hammerless guns (Needham's 1874 patent is the first recorded).

In lever-cocking guns, such as the original 1871 Murcott 'Mousetrap' and variants on the theme like the Woodward 'Automatic' and the Gibbs & Pitt, the tumblers are cocked when the lever is pushed to open the gun.

In barrel-cocking guns such as the 1875 Anson & Deeley, the leverage of the falling barrels, when released by the operating-lever, acts to cock the tumblers. In other barrel-cocking designs (such as the original 1883 Holland & Holland 'Royal' and the Perkes patent used by Boss for their early hammerless guns), the barrels may cock one lock by the falling of the barrels and the other by the closing motion, though the later 'Royal' abandoned this complication in favour of cocking both locks via the drop of the barrels.

Technically, the Beesley/Purdey is a 'spring-cocked' action, as the mainspring cocks the lock. The mainspring itself is placed under tension by the raising of the barrels. The main distinction however between conventional cocking actions is the role of the opening lever and whether it is used to cock the tumblers or simply to open the gun.

Lever-cocked guns require more effort in the operation of that lever and were generally dropped in favour of barrel-cocking, but barrel-cocking guns are sometimes criticised for the 'drag' experienced in the falling of the barrels as they cock the tumblers. Spring-cocked guns have been criticised for the extra effort needed to close them, though lauded for their ease of opening.

This Grant trigger-plate action gun is cocked when the side-lever is pressed.

exclusive hold over the 'best' guns moniker – guns of the finest quality will regularly be encountered of boxlock or back-action sidelock form, as indeed they will be found as hammer and trigger-plate action guns. When considering a gun from any era – look for the quality in the piece, for this will usually be of more significance than the design.

One has to sympathise with W.W. Greener a little when considering the eventual out-muscling of the boxlock for 'best gun' status. To Greener's credit, his best boxlocks are fantastic examples of ingenuity, engineering and craftsmanship. Perhaps these factors were part of the reason for their downfall, only Greener specialists could cope with the demands placed on the maker and regulator of the 'Unique' and 'Facile Princeps' actions. To quote the man himself on the G-grade gun (his best boxlock):

'...requires the most careful adjustment and although the parts are few and most simple, to ensure perfect working, the utmost precision is necessary in centring and shaping the various limbs'.

As 'London pattern' sidelocks became the 'best' norm, the Anson & Deeley pattern boxlock found more general favour with makers than the more complex Greener designs for use in lower-grade guns. Greener's actions developed a rather unfair reputation for unreliability that probably stems from gunsmiths unfamiliar with the design trying to service, regulate or adjust them – and wrecking them as a result.

The years have certainly proved that good quality sidelocks and boxlocks (and indeed hammer guns) will last almost indefinitely if treated with a little care and basic routine cleaning. All will perform admirably in the field today and will continue to do so for years. There is no reason why you cannot buy a gun, of any of the types described above, from the 1870s or 1880s and use it every season for the rest of your life and pass it on to your son with every expectation of him getting the same use out of it.

Hammer versus Hammerless Guns

Burrard and Greener may not have agreed on their choice of the best gun-lock but they did agree that hammerless guns were preferable to hammer guns, although Burrard had nothing to say against hammer guns other than to note (in 1930) that *'they are seldom made now except in the very cheapest grades'*. By the time Burrard was writing, the hammer-vs-hammerless debate must have seemed academic, such was the dominance of the hammerless gun

by then. Their shared view, that hammer guns had been surpassed by the newer hammerless designs, was very much in-line with other eminent contributors to the debate.

H.A.A. Thorn, writing as 'Charles Lancaster' in *The Art of Shooting,* appears to have viewed the residual affection for hammer guns during the transitional period as an indication of conservatism and habit on the part of established shooters. *'There were arguments reasonably urged against the hammerless, but perhaps the most powerful, though with the least to say for itself, was the instinctive feeling that without hammers the gun did not look right'.* He goes on: *'The hammer breechloader is still occasionally made but the verdict of contemporary opinion overwhelmingly favours the hammerless system'.*

Here is Dr J. H. Walsh, long-time editor of *The Field* (1857-1888), writer (under the *nom de plume* 'Stonehenge') and great contributor to the promotion of 19th century gun improvement (notably in organizing 'Field Trials' to assess the merits of the new inventions – most famously 'muzzle-loaders vs breech-loaders' (1858-1859), 'choke-bored vs true-cylinder' guns (1875-1876), and the 'Field' Rifle Trials of 1883) on the subject: *'The hammerless gun is the superior in point of safety and efficiency... the hammerless gun is, I think, to be preferred'.*

W.W. Greener, again, writing in 1888 in *Modern Shotguns,* was in more of a mood to set out the case for hammerless guns. At that time, customers would still be debating whether a new hammerless gun was necessary. Greener, ever the salesman, precedes the customary lauding of his own hammerless 'Facile Princeps' gun with no less than seven reasons why (his) hammerless guns are superior:

1st They are safer

2nd They are quicker

3rd Caeteris paribus, they are stronger

4th They are less liable to damage

5th They are more reliable

6th They are less complex

7th They are more handy

*They **may be** more durable, more economical and more beautiful.*

In the matter of complexity, hammerless sidelocks are no less complex than hammer guns (though remember Greener was obliquely referring to his own boxlock with only three moving parts) and in the matter of economy

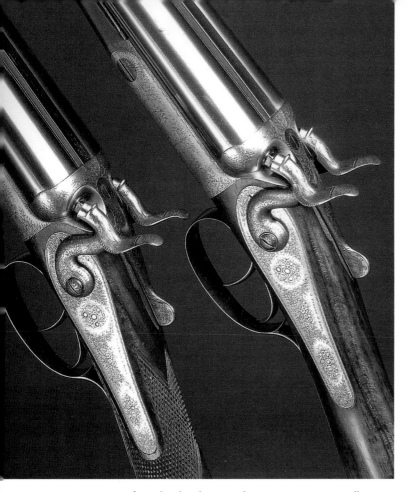

A pair of Purdey back-action hammer guns originally belonging to the Marquis of Ripon. These have steel barrels and non-rebounding locks and are choked 'extra full' in each barrel. The barrels are 'Whitworth' steel replacements, fitted in 1909.

Greener appears to contradict himself on the next page when he states '...*each hammerless gun he* (Greener) *makes costs him more than the same grade of hammered guns...*'

As a user of hammer and hammerless guns of all types for a wide variety of shooting, I would also take issue with Greener on points 1, 3, 4, and 5. In fact, in all shooting situations except high-volume driven birds, where speed of loading is paramount, I find little to choose between the two. Indeed, if asked to single out one of the three (boxlock, hammerless sidelock and hammer gun) to remove from my collection, I would have to select the boxlock: aware as I am of its qualities.

It is a matter of historical fact that the hammerless gun overtook the hammer gun as the weapon of choice for the British shooting public, much as the over-and-under replaced the side-by-side a hundred years later. However, some noted die-hards continued to have hammer guns made well into the 20th century, notably King George V and the Marquis of Ripon. Hammer guns also retained a following

in 'live pigeon' shooting circles well into the 1920s. (Live pigeon shooting was banned in England in 1921, under the Captive Birds Act of that year, but continued on the Continent).

Reconsidering the Qualities of Hammer Guns

Gunmakers' records illustrate the leisurely manner in which hammerless guns took over from hammer guns in terms of customer demand in the transitional period: in 1878 Holland & Holland produced only six hammerless guns and 158 hammer guns. By 1880, hammerless guns outnumbered hammer guns by 111 to 89, but 1881 saw a reversal with hammer guns outselling hammerless in three of the next four years. Only in 1885 did the decline in hammer guns collect momentum.

Purdey records show 208 hammer guns against 33 hammerless for 1880 (the first year of the Beesley patent hammerless action). Hammerless guns did not overtake hammer guns in the order books until 1886. Donald Dallas in *Purdey: the Definitive History* (2000) tells it thus:

'The switch to hammer guns was relatively pedestrian. The advantages of the hammerless gun over the hammer gun were not so great that customers had to have the new design. The change in the 1880s was very gradual and it was only with the advent of the ejector gun in the late 1880s that hammerless guns were ordered with greater frequency... even then a large number of hammer guns continued to be built right up to World War I.'

Webley & Scott's catalogue of 1914 offers hammer guns in all qualities. Their W & C Scott 'Premier' hammer pigeon gun is offered as a 'best' gun at $336 with Damascus or $376 with Whitworth steel tubes. The same grade 'Premier' hammerless pigeon gun is somewhat more expensive at $406 with Damascus, or $445 with Whitworth tubes. It is interesting to note that the hammer guns are offered in 'best' quality as pigeon guns only. The cheapest hammer gun in the catalogue is a plain back-action gun retailing at only $26. Their cheapest Anson & Deeley gun is $65. The catalogue is a wholesale issue for the gun trade, so retail prices would be higher at point of sale.

By 1934, the price of hammer guns on the second-hand market had fallen dramatically in comparison with hammerless guns. An advertisement in *Game & Gun* of that year by C.B Vaughan of the Strand, London offers a used Purdey 16-bore hammerless sidelock ejector for $277 while his Purdey hammer ejector is just $86 and the Purdey

My current favorite bar-action hammer gun by J. Thompson. In the 2006/7 season I used this gun almost weekly on driven pheasant days, relegating my Purdey sidelock to the status of back-up gun.

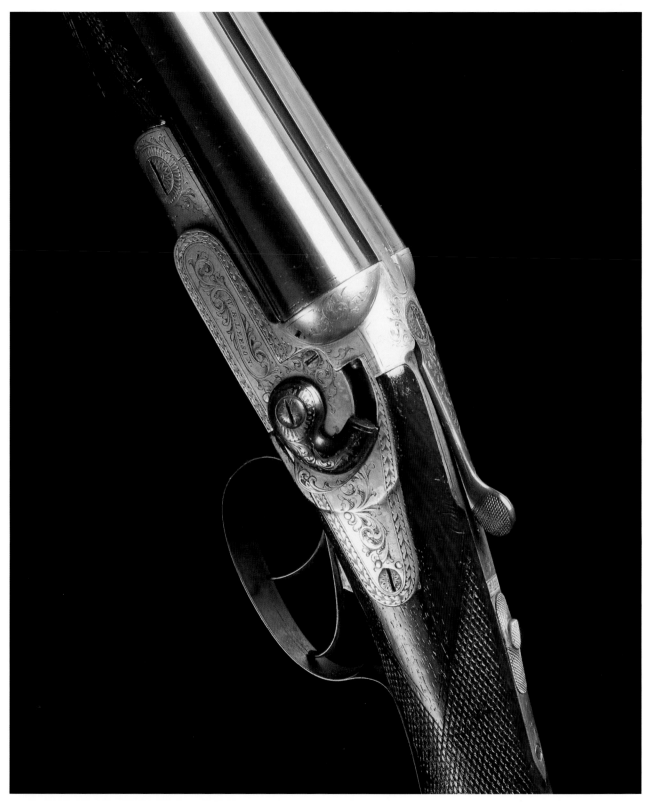

The semi-hammerless gun was not a great success. But this Bentley 1884 patent offers a fascinating insight into developmental thinking in the 1880s.

hammer non-ejector on his books is a mere $50. At the time a new Holland and Holland 'Royal' cost $606, their 'Dominion' model cost $252 and Alex Martin advertised guns ranging from $38–$380. The 'best' used Purdey hammer gun cost only a little more than the cheapest gun Alex Martin offered new.

Resurgence in appreciation of the hammer gun among more recent authors is interesting and offers us the chance to reconsider their relative merits after several lifetimes of comparison; Gough Thomas (GT Garwood) wrote on the subject in *The Shooting Times* in 1967: '*I understand that there is a lively and possibly an increasing demand for undeniable old thoroughbreds in good condition, which may fetch $220-$250*'. Compare this with the price of a new Purdey hammerless ejector at the time of $4,120. The type of hammer gun 'GT' was writing about would have cost $370 new, in 1900. Today a Purdey hammer-ejector in top condition will cost around $23,700 at auction but $5,900 will still buy you a very nice hammer gun by a 'best' maker.

Three decades later, American authors Cyril Adams and Robert Braden's appreciation of the hammer gun is clearly outlined in their excellent retrospective *Lock Stock & Barrel* (1996). They even make the case that the hammer gun, rather than the bar action hammerless sidelock, embodies the perfection of sporting gun design. I quote:

'*More and more discerning sportsmen and a few competitive live pigeon shooters learned that the past 120 years have not produced a shotgun that has all the intrinsic qualities of the hammer gun*'.

Their argument is interestingly reminiscent of Richard Arnold's earlier (1950s) fond contemplation of the muzzle-loader and its shooting qualities in his *The Shooter's Handbook*.

Adams & Braden go on to argue that hammer guns outperform hammerless designs in the following categories:

- Balance (weight is between the hands as the action is more compact)

- Size of action (hammer guns can have smaller actions yet remain strong)

- Accommodation of longer barrels

- Simplicity and durability

- Ease of opening and closing (there is no barrel-cocking system to impede operations)

They clearly appreciate the qualities of the hammer gun and its performance as an effective tool for the dispatching of game but also show their sensitivity to the artistry of the gunmaker: '*it is also reasonable to conclude that best-quality hammer guns are the most elegant and beautifully proportioned guns ever made*'. It is refreshing to note the aesthetic argument spanning the generations of shooting writers; Burrard and Greener had also, almost reluctantly, added this to their evaluations of a gun's merits. Here we have a 'full set' – Burrard found hammerless sidelocks the most beautiful, Greener the boxlock (admittedly his own version) and Adams & Braden the hammer gun. Though they come to different conclusions, they all show their appreciation for the intrinsic beauty of quality guns.

Vintage guns competing with new ones in the market place

Another modern author, Mike George, focuses on purchasing issues for the modern Shot: choosing between the traditional side-by-side and the modern over-and-under. He discounts old guns and new English side-by-sides as choices for most people, perhaps understandably dismissing the latter as '*a few hand-made exotics, costing at least as much as the average man earns in a year*' and failing to address the former as an option at all. In his advice to readers, rather than worry about the action type and its theoretical merits (he has seen enough in 100 years of efficient boxlock, trigger-plate action and sidelock performance to move on to more fundamental issues – they clearly all work) he analyses in some detail the merits of modern factory-made over-and-under guns.

In *The Shotgun Handbook* of 1998 George argues '*I have tried to get rid of the idea that 'hand-made is best' because in our terms it isn't – we need well-made guns at prices we can afford. Our guns, like the cars we drive, are made by machine tools in modern factories.*' He concludes his history of the growth of imported guns to the UK and the rise to dominance of the European, American or Japanese over-and-under by saying: '*the average shooter with between $830 and $8,300 to spend chooses an import because, if he wants a new gun, he has no other choice*'. The flaw in his argument lies in that part of the final sentence.

George argues the case for the imported gun well and his account of the demise of the British gun trade because of its refusal to move with the times is sadly accurate and reflects the realities of our modern world. But why spend your money on a new gun? You have absolutely nothing to gain by doing so. At the start of the 21st century we have a huge variety on offer to us as shooters and the decision for many of us is more personal than merely practical.

At the end of the 19th century and the start of the 20th the focus was on technology and invention; developments pointed towards the perfection of a shooting system. Everybody was inventing improvements, until the inventions began to have little practical impact on the shooter. Many would argue that after about 1900, there was little left to do to fundamentally improve the sporting shotgun. Browning aficionados may argue about this but, although technically brilliant, can one really argue that the repeating shotgun is more efficient, more elegant and more suitable as a weapon for dispatching driven game? Certainly it is cheaper and easier to manufacture but that is not the point.

Top: a Boss bar-action sidelock with intercepting safety indicated. When the trigger was pulled, the intercepting safety was pushed clear and the tumbler allowed to fall. Bottom: a Purdey bar-action sidelock with intercepting safety indicated. Should the sear be jarred out of the bent, allowing the tumbler to fall without the trigger being pulled, the intercepting safety will prevent it from reaching the striker.

Safety Devices

'*Hammerless (guns) are as safe as guns can be, provided they are fitted with a reliable intercepting block that is always between the hammers and the cartridge...*'

Sir Ralph Payne-Gallwey, 1890.

Hammer guns are easy to see as cocked and ready to fire or 'at rest' and incapable of discharging a cartridge; the big, visible, hammers act as a safety device and a cocking indicator. When hammerless guns emerged many shooters were wary as to their safety, as when closed, the tumblers are automatically cocked.

Some early safety catches merely prevented the trigger from being pulled, but this did not prevent the sear being jarred out of the bent by a knock, a fall, or a dirty or faulty bent and sear not connecting properly. In such circumstances, it was possible for the tumbler to fall independent of the triggers – and fire the gun. Inventors quickly began to rectify this problem by working on systems that would physically block the fall of the tumbler *unless the trigger was pulled.* A good example of this work is the Scott & Baker 1878 back-action, which incorporates an

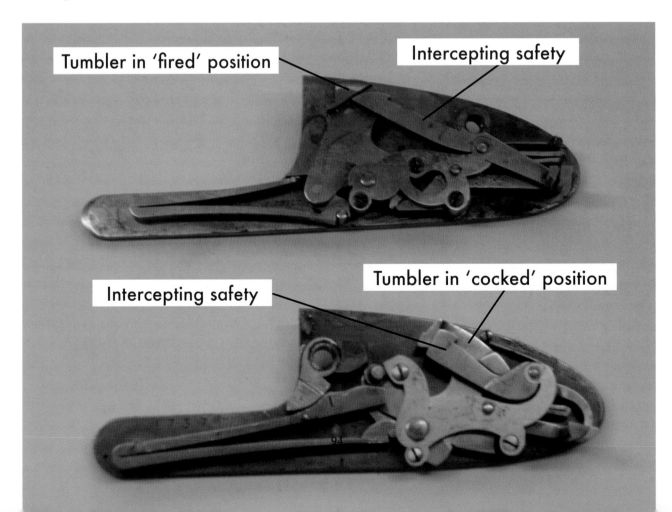

Tumbler in 'fired' position

Intercepting safety

Intercepting safety

Tumbler in 'cocked' position

Guns that can deceive the unwary

Many British guns are styled externally to resemble the conventional bar-action sidelock, despite the lock work being very different on the inside. Careful inspection is essential.

 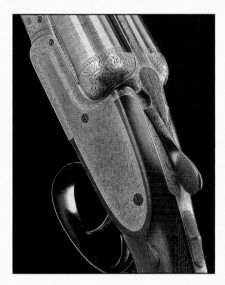 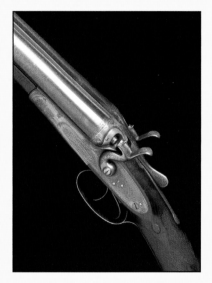

The Cogswell & Harrison gun on the left looks like a sidelock but is actually a side-plated boxlock. The Dickson in the middle looks like a sidelock but is actually a round-action trigger-plate lock and the Akrill hammer gun on the right looks like a bar-action but is actually a back-action.

'intercepting safety bolt'. This blocks the fall of the tumbler and prevents it reaching the striker. Pulling the trigger moves the safety bolt out of the way and allows the tumbler to fall fully and therefore, the gun to fire.

Most hammerless guns also have a device to lock the triggers, so that they are not inadvertently pulled by clothing or vegetation. The location of the operating stud for placing the safety 'on' or 'off' is generally on the top-strap of a British double gun. It is generally pushed forward to disengage and pulled back to re-engage. The word 'SAFE' is often carved, or inlet in gold lettering, and is obscured when the safety is in the 'off' position. Some, like the 1879 MacNaughton have a lever in the top-strap, which is pushed left or right to engage or disengage.

Greener boxlock guns are often encountered with a safety on one of the side panels of the stock. Greener believed this allowed the hand of the stock to be kept stronger and he preferred the aesthetics. He also believed the safety should not be automatic, though he would make it so for an extra $5.

Occasionally one will find guns with safety devices in the butt-plate, which disengage as the gun is put to the shoulder. Others are operated by a rod standing proud of the hand of the stock, which disengages the safety bolt as the hand is gripped for shooting. These and others proved unpopular and by the mid-1870s the top-strap mounted safety catch was in the ascendant.

When do you need a Certificate and when do you not?

It is not universally known that British collectors may own certain guns without a Shotgun Certificate (SGC) or a Fire-arms Certificate (FAC). The rules are not entirely straight-forward but here is a basic guide to current requirements. It is noteworthy that laws are prone to change. Before purchasing any gun it is a good idea to seek the advice of your local Firearms Enquiry Team. A good start is the Metropolitan Police SO19 Team in Hendon: www.metpolice.uk/firearms-enquiries/

There are three main reasons for people owning weapons 'off ticket':

• Because ammunition is no longer available for the gun and it is considered an 'Obsolete Calibre' and kept as a curio.

• Because the gun is 'Antique' and is not intended for use.

• Because the gun has been deactivated and is incapable of being fired.

Muzzle Loaders

If you have a muzzle-loading shotgun of pre-WWII vintage it can be kept without any licence, provided you do not attempt to use it or own ammunition for it. As soon as you intend to use the gun, a Shotgun Certificate will be required.

Breech-loading Rim-fires

If the calibre exceeds .23" rim-fire calibre (excluding 9mm), no licence is required as long as the gun is not intended for use. It should be noted therefore that the common .22 rim-fire rifle and the 9mm Flobert 'garden gun' rim-fire shot cartridge require FAC and SGC respectively. Many 'rook rifles', such as the old .360 are now classed as obsolete. However, be careful to check such guns because many have been modified to take .22 rifle or .410 shotgun ammunition.

Breech-loading guns or rifles

If the weapon is pre WW2 and not a centre-fire or rim-fire system, then it is exempt from certification (this applies to needle-fire and pin-fire guns for example).

Obsolete shotgun bore sizes

The following bore-size shotguns may be purchased and kept without a SGC as long as they were made before WW2 and are not intended for use: 24-bore, 32-bore, 10-bore (some chamber lengths notwithstanding), or larger, such as the common 8-bore and 4-bore wildfowling guns, including breech loaders. Be aware that if you are in possession of ammunition for such a gun held 'off ticket', you probably invalidate your claim to have no intention of using it.

8-bores like this Holland & Holland can be kept as curios without a Shot Gun Licence or Firearms Certificate.

Deactivated Guns

The 1988 amendments to the Firearms Act of 1968 provide for the 'deactivation' of firearms to make them incapable of use. They then become 'objects' and are not considered firearms or shotguns under the law. A deactivated gun must carry a deactivation mark from the proof house and a deactivation certificate.

Ammunition

No ammunition is considered obsolete. To own rifle ammunition, you need a FAC and permission to have the quantity and calibre specified. You do not need a SGC to possess shotgun ammunition but you need one to buy it.

Modern 'Antiques'

Any gun of 'modern manufacture' (generally accepted as after WWII) needs a licence regardless of the above. Therefore, a modern Pedersoli muzzle-loader needs to be put on a SGC whereas a vintage Manton muzzle-loader of the same bore does not.

By 1910 the shotgun had become the perfect weapon for the job for which it was intended. 'Improvements' have largely been made in the manufacturing process, as the mass-market gun has stopped being a hand-made article and become a piece of precision engineering from an automated production line (actually BSA were doing this in the 1920s).

This happened with cars, but it took a little under a century longer to get to the point where, to use an oft-quoted cliché, 'there are no bad cars any more'. The sensible modern choice is the Ford Mondeo. It is faultless – no really, it will top 100mph with ease, is economical to run and service and is very reliable. It will transport the family in comfort, at speed, in safety and with economy for years and even holds a good resale value. But not everyone drives a Mondeo. Why not? It is, after all, the logical choice.

Those with the 'Mondeo Man' mentality towards guns shoot a Beretta or a Browning, a Lincoln or a Miroku when out in pursuit of game. Those of us who choose an older gun do so for the same reason that keen drivers select a TVR or a vintage Daimler. We know they don't add up in mathematical terms but the quality of our experience makes it worthwhile. We operate on another plane and it makes our lives richer.

Try choosing a gun to take out that you know is not the easiest to master, to clean, to be seen with. It will add another dimension to your shooting. Game shooting is about more than facing your bird with your tool in your hand (apologies to Derek & Clive) – yet your choice of 'tool' is a mark of your respect for your quarry and your personal involvement with the contemplation and engagement of the chase.

Honour the pheasant by matching his flight with your skill and a historically interesting gun. You will surpass the simple slaying of a creature with a complex mixture of nostalgia, empathy and challenge. You meet him with equipment his forefathers and yours recognized in battle. You need no more if your skill is up to the challenge. I guarantee every kill will be more satisfying, more memorable, less industrial in nature.

Boxlock or sidelock, hammer gun or hammerless matters not. Select your gun or guns according to the qualities and charms you find in them. You will notice this book shamelessly mixes practical evaluation and analysis with arbitrary selection based on subjective matters such as appreciation of those indefinable qualities 'charm' or 'style'. This is how it should be – otherwise we would all be shooting with a plastic stocked semi-auto.

A lot of the time we spend shooting is actually spent waiting. We wait with gun in hand. The gun you have in your hand is part of the value of the experience. I stand at my peg, sit in my pigeon hide or crouch in the undergrowth waiting for an evening flight, with a gun in my hand that pleases me just to look at, to hold it and to contemplate shooting it. Even If I do not discharge my gun, all this I get. No simple tool.

In morning sunshine or the fading light of dusk, the craftsmanship, the engraving, the balance of old guns is there to cause one to ponder the men who did the work so lovingly, so meticulously, so terribly well. You don't get that from looking at the alloy action of a Browning Cynergy.

So much for reflection on sensibility; each to their own. My sentiments reflect essential elements of my own approach to shooting and you will either find we have common ground or disagree fundamentally.

The ubiquitous Beretta Silver Pigeon. Reliable, affordable and seen on shoots everywhere. As a tool for shooting it has few peers in its price range.

Other technical matters

Ejectors

The history of the ejector is interesting. Inventors throughout the second half of the 19th century who were working on guns seem to have perfected the various working parts of breechloaders in sequence according to importance. First a secure system for fixing the breech to the action face was required – Jones managed this in 1859 and Purdey perfected it in 1863. Detonating was perfected in hammer guns with improvements on Stanton's rebounding lock by 1867. Hammerless guns were made reliable and effective in

1875 with the Anson & Deeley boxlock and the hammerless sidelock reached perfection with the 1880 Beesley patent.

The next challenge was to increase speed of loading. Snap-action operating systems such as the Purdey bolt and Scott spindle combined with the top-lever helped the shooter open and close the gun rapidly but he still lost time removing the cartridges with his fingers. Remember, the lucky sportsman of 1880 shooting with a new Purdey sidelock given to him for his fiftieth birthday could have started his shooting career on his twentieth birthday with the gift of the latest percussion muzzle-loader. Enormous improvements had been made in a very short period of time but still the sportsman demanded more. Gunmakers got to work on automatic ejectors.

Needham made a workable ejector in 1874 with a design housed in the body of the action. This required a rather unsightly projection to be fitted immediately ahead of the trigger guard on the underside of the action. However, Needham maintained the patent rights for the full 14 years, spurring others to come up with alternatives, and they, including Greener, in 1881, patented derivatives of the design, based inside the action and powered by the mainspring. Other patents, Thomas Perkes' for example (1878), housed the ejector work in the forend. Deeley produced an ejector in 1886 in the form of a separate working lock, housed in the forend, for use with the boxlock. This is commonly encountered in Westley Richards guns. It could also be adapted for sidelock application but was rather more complicated than the system that was to supersede it.

The Perkes 'over-centre' principle, patent of 1878, formed the basis of possibly the most simple and most favoured ejector system; known commonly as the 'Southgate Ejector' or sometimes the 'Holland Ejector'.

Southgate patented an ejector in 1889 with follow-ups in 1890 and 1893 and Henry Holland patented his variation on the 'over-centre' theme in 1893. However, Donald Dallas in *Holland & Holland The Royal Gunmaker* notes that the development of the 'over-centre' ejector and its culmination into the classic 'Southgate' is more accurately

credited to Perkes for his initial over-centre patent and that great 'Inventor to the London Trade' Frederick Beesley with his 1889 patent. Dallas reminds us *'The ejector should be more correctly described as the Perkes/Beesley rather than the Holland/Southgate'.* The over-centre principal does have its detractors though. The main criticism is that it is a friction-fit system and is therefore prone to wear and require regulation and repair after heavy use. Also, being reliant on 'V' springs, it ceases to work when the spring fails. However, these objections carry little practical merit.

Major Burrard had also, earlier, credited Beesley with perfecting this system in *The Modern Shotgun* of 1931. It is indicative of the times that so many of the inventions we know today and identify with certain gunmakers were actually the work of others. Beesley is a good case in point – he invented the 'Purdey Sidelock' and inspired the 'Holland Ejector', two of the most famous and copied systems ever devised for use in a sporting shotgun – yet neither carry his name in common parlance.

The ejector systems described earlier are based on a simple mechanical principle; they are easier to regulate and more reliable than many of the other systems tried by various gunmakers while these patents still protected the patentees.

The other notable ejector system, working from an entirely different angle of thought is the Boss system. It uses a coil spring in the forend and is instantly recognizable by the fact that the extractors lift the unfired cartridges significantly higher than other systems. The basic difference between ejecting the cartridge and simply holding it proud

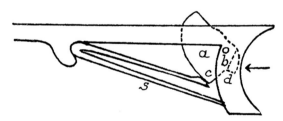

The forend-housed Holland ejector is simple, robust and reliable. It works on the 'over-centre' principle and has been used by gunmakers for well over 100 years.

Key Ejector Patents								
1874	1878	1881	1886	1888	1889	1891	1893	1897
Need-ham Ejector	Perkes Patent	Greener Self-acting ejector	Deeley Ejector	Wem Ejector (Purdey)	South-gate Patent Beesley Patent	Holland 'AB' Ejector	Holland Patent	Boss Patent

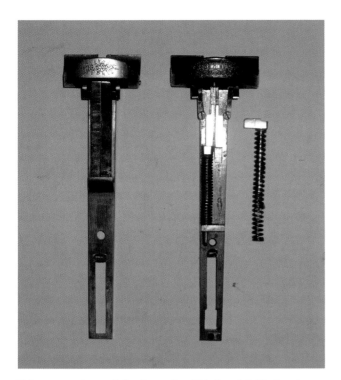

This comparison of the forends of Holland & Holland (left) and Boss (right), both side-by-side sidelock ejectors, illustrates the use of leaf springs in the former and coil springs in the latter (right spring detached to show full extension).

of the breech end is the force exerted by the coil spring. When fired, the spring's energy produces a 'flick' and the slide moves fast enough to eject the empty. When unfired, the slide simply moves upward under the more gentle force of the spring. The Boss system is very reliable, easy to regulate and even works when the coil spring is broken.

Guns made as ejectors or converted to ejector during the 1880s and 1890s require careful evaluation of the ejector mechanism to make sure it is working efficiently. When they go wrong they can be very tricky to regulate and make reliable. Some were just not good designs and fell into disuse as a result – but they still lurk in the insides of many vintage guns waiting to catch out an unwary purchaser. In *The British Shotgun 1871-1890*, Crudgington and Baker illustrate no fewer than 93 different ejector mechanisms patented during the period in question.

Single Triggers

James Templeman patented a single trigger as long ago as 1789 and subsequent attempts at single triggers were made throughout the history of gunmaking, Rigby for example

used them in pistols from the 1860s, but it was not until the 1890s, when everything else about the breechloader was more or less what it is today, that the single trigger development really began to attract serious attention and at this time the first reliable mechanisms began to arrive at the patent office.

The basic problem with single triggers is the fact that when a shotgun is discharged and recoil delivered to the shooter, a second 'involuntary' pull of the trigger occurs that the shooter is unaware of. This is clearly a problem because it causes the shooter to fire both barrels when he only intends to discharge one. Single trigger developers had to overcome this 'involuntary pull' issue before they could make any single trigger usable.

Two distinct avenues of development were followed by British inventors tackling this problem:

- Three-pull systems
- Delay systems

In the 'three-pull' systems the principle was that the mechanism required three pulls on the trigger to fire two barrels – the second pull did nothing, only the first and third pull effecting the discharge of the gun. This is typified by the 'Boss' system patented by John Robertson in 1894. Public trials in that year proved the reliability of the new Boss single trigger. This success led to further public trials in 1896, met with universal press acclaim, and a further test in *The Field* in 1897, which cemented its reputation for reliability.

Boss made single trigger guns, based on patent No. 22894, from 1894 onwards. It was developed and improved until 1905 and became something of a Boss trademark. Such was the success of the system that it is commonly found in guns by other 'best' makers, where single triggers have been added some time after their original delivery.

In the 'delay' systems a totally different approach was taken. The system required one pull to fire the first lock and then blocked the mechanism for the second, involuntary, pull so that when it occurred nothing happened. The second conscious pull fired the second lock.

The 'Holland' single trigger patented in 1897 by Henry Holland and Thomas Woodward is a good example of this type of trigger. Westley Richards also had a reliable single trigger by 1887.

Teasdale-Buckell adopted an unconventional trial of the Holland single trigger, which he recounted in

'*Experts on Guns and Shooting*' in 1900. After putting the gun through its paces with a high number of cartridges at repeated clay targets without any mishap: '... *by way of seeing whether moth and rust were likely to corrupt, two single trigger actions were placed in water and a shovel full of ashes fresh from the fire thrown in and stirred up. The actions, of course came out covered with big and small ashes. In this state they were returned to their respective guns and we began firing again... Holland's single trigger stood the test triumphantly*'. I'm not sure this approach to shooting journalism would be tolerated by gunmakers these days!

Both the Boss and Holland single triggers are used in their guns today and are much copied. Modern single triggers like the Miller are crude but effective because they have few moving parts. More sophisticated modern mechanisms such as the continentally inspired Purdey bob-weight system are reputedly also very efficient.

Other issues

We have explored the evaluations of major writers on the merits of the many developments in gun design that took place during our period of interest. Be guided by them if you will or allow your own evaluation to hold sway. There really are no rules, just useful things to know. We shall now consider the various offerings likely to avail themselves to us and discuss their merits or limitations.

It is time to consider what one may purchase in order to enter this world of the practical eccentric. We have established the eccentricity and it will either appeal or it

What could you buy second-hand in 1926?
Game & Gun advertisements of the time give a fascinating insight into the comparative prices of second-hand guns offered for sale in London.

Purdey Sidelock Ejector $252
Holland 'Royal' Ejector $243
Boss Sidelock Ejector $243

New guns in the same publication include:

WJ Jeffery sidelock hammerless ejector $155
Greener hammerless ejector $252
Grant & Lang hammerless sidelock ejector $180
Army & Navy hammerless sidelock ejector $170

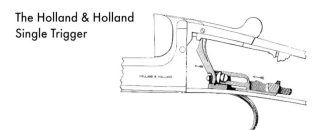

The Holland & Holland Single Trigger

will not. If it does, you are in good company. Now we will consider some practical issues.

It is easy to drool over the glossy pages of guides, histories and brochures illustrating the products of Purdey, Holland & Holland, Boss and any number of 'best guns' offered as new or pre-owned. These are often the gold-encrusted preserve of the investment-minded collector or the very wealthy, they require the bank balance that goes with a City salary and they are fabulous pieces. But, may I say, they are the obvious, the easy, even the lazy way into the world of fine guns.

Do not despair that you cannot afford a mint pair of Boss sidelocks from the most favoured years of manufacture. Look leftfield and there is still a world of accessible wonder to discover. You may only earn an average salary but you *can* have a Purdey.

The world of gun collectors is inhabited by some very wealthy individuals who are particular about defining the, supposedly, most desirable features of a sporting gun. These features have entered the psyche of the gun trade and have become the 'conventional wisdom'. It is easier to think what others tell you than work things out for yourself when making a purchase.

Let us imagine for a moment that you just got your City bonus and have the taste to choose a vintage English gun as opposed to a non-English over-and-under covered in gold ducks or one of those attractive but not really up-to-the-mark Italian pieces. You want to buy wisely and have 'the thing' for those formal driven days to which you are partial.

The conventional wisdom of the day suggests the following:

You need to have a sporting gun (or better-still, a pair) made by Boss, Woodward, Holland & Holland or Purdey. It should be a sidelock ejector with a top-lever, 28"-30" barrels of chopper-lump construction, French walnut stock and splinter forend. It should be from the 'Golden Age' (1920-1940), totally original, close to original barrel measurements and contained in an original oak & leather maker's case.

Was there ever a 'Golden Age' of gun-making?

Writers and gun dealers often talk of a 'golden age'; a time in which the very best gunmaking and the very best guns coincided in an era that has never been surpassed. The period covering 1920-1940 is generally mentioned. I'm not convinced.

A gunmaking friend told me he believed the very best craftsmen from any period would be able to match the standards of any other. I bow to his authority on this for perhaps it takes a craftsman to really understand another's work. Men who spend their whole lives shaping wood and metal surely appreciate best work when they see it. It is interesting that few actual gunsmiths write on the subject. I have had it put to me that most writers are only expressing an opinion and the worth of that opinion is often questionable. 'A lot of gun dealers pass out quotes as if they were gunmakers — but they are not. Any fool can pick holes in a gun' as one bone fide London gunmaker put it to me: his temperature visibly rising as we broached the subject of critics. Who are we to judge a man's work – we who cannot do it or hope to understand the input, precision and artistry involved?

However, my uncultured eye notes a subtle difference in the lines of a Purdey sidelock from 1890 and one from 1955 – why should this be, when did the changes happen? It has been described to me as a series of 'Chinese –whispers'. An action filer would teach his apprentice his skills but one eye and one pair of hands differs subtly from the next and the lines change slightly in time as the generations take over from one another; even as time passes – look at your signature from when you were twenty and compare it with today's – the years will have left their mark. It is the same signature – but not quite.

If the 1930s are seen as a golden era, perhaps it is because there are a lot of best guns in good condition from that period. This decade was interrupted by a war that ended the lives of many owners and the post-war period saw a decline is shooting as the old order was replaced with the new. The great days of Edwardian shooting parties were over. Guns were less likely to be expected to fire 500 shots a day five days a week, four months of the year. Many were obviously used sparingly or mothballed to emerge pristine on the market years later. We still see them. Victorian and Edwardian guns are more likely to be worn out, or damaged

and repaired from trips to the colonies. To my eye the lines of an 1885 sidelock and the subtle fine engraving of the period is hard to better. The quality of some 1870s hammer guns has to be seen to be believed.

Perhaps the 1970s was the antithesis of a golden era. 'The decade that taste forgot' was not the highpoint for even the finest gunmaking firms. Rapid inflation meant many new guns were being sold at prices two-thirds of their market value. The price quoted at order was overtaken many fold during the time it took to produce. Gunmakers were under pressure to make guns faster. It can't have helped; and guns of this period do lack a degree of finesse, in the opinion of many observers. Comparison is enlightening, but the reader will make up his own mind on the subject.

Essentially the style of gun you are looking for is an example of the perfection of the sidelock ejector, reached just before 1900. By this time locks, safety devices, opening and closing mechanisms, ejectors, visual style and proportion were what we recognise today.

The Beesley action of January 1880 patent is the Purdey sidelock you will buy if you walk into South Audley Street today (Purdey bought the rights to the design but

Best sidelocks with their case and accessories and some historical provenance are the top of the collector's list. This Purdey belonged to Prince Frederick Duleep Singh, son of Prince Victor Duleep Singh, a legend of the Victorian and Edwardian shooting scene.

let's credit Beesley for his invention here). The Holland & Holland Royal of 1884 latterly lost its old 'leg of mutton' lock shape in favour of a more conventional style, now more closely resembling the Purdey and the (also re-designed) Boss.

The ejector had reached the stage of solid reliability with Southgate's patent of 1889 and its subsequent improvements. Although single triggers had a little work still to be done, these are not universally liked and remain an option rather than a standard.

So the 'ideal', coveted by the wealthy collector or user of English sporting guns is a vintage version of the current 'best' gun. Had I the money, I would certainly buy one; for they are much nicer than new ones in my opinion and I should rather have a 1904 Purdey in original condition than a 2004 version. I say this with no disrespect to current Purdey output, which is of fabulous quality.

The fact that $40,000 will buy you as beautiful a gun as it is possible to imagine, from the turn of the century, makes the $100,000+ cost of a new gun seem like money unwisely spent, if money is an issue. This logic pervades across the board; you will get a better gun buying second-hand than you will by buying a new gun on a pound-for-pound basis.

So much for our friend in the City with his $40,000 and his perfect vintage sidelock. My girlfriend and your wife and, probably, our incomes conspire to prevent us from considering such a magical purchase. So let us consider what we need to understand about our selection of a gun. What features are fashionable now (and costing a premium) and what features actually contribute to the shooting quality of a gun? What do we intend to use each gun for and what will it bring to our lives that other guns do not?

The statements which follow are commonly made and they support the conventional wisdom that dominates the pricing structure of guns for sale. We shall consider them from the point of view of the practical eccentric and see where the truth lies.

'Sidelocks should be stocked to the fences'

Why, and who says so? Most gun books will tell you this is a fact; Chris Austyn in *Modern Sporting Guns* for example. The horns of the stock on 'best' guns nowadays go right up to the fences in the convention set following the success of the Beesley/Purdey of 1880. Many older sidelocks were made in a period before this convention had taken hold and the feature is no indication of inferior quality.

Early Boss and Holland & Holland and William Powell hammerless sidelocks were not stocked to the fences and neither were many other designs retailed in the late 19th and early 20th centuries. All other things being equal, a gun not stocked to the fences will cost you less. Will it make any difference to your enjoyment of the gun, its elegance and charm or indeed its shooting qualities? No. This is another example of how the conventions and prejudices of the general gun-buying public can be exploited by those who judge guns by other merits.

'Ejectors are important'

If you are shooting driven pheasants and partridges then this may be seen as a reasonable statement and, though I often shoot in such circumstances with a hammer non-ejector, I would say ejectors are for most people necessary for driven bird shooting.

How much of your shooting is driven partridge or pheasant? I suspect that most of us shoot at least as often at non-driven birds as at driven ones. In these circumstances, birds will usually come less frequently and the need to 'speed load' simply does not exist.

There is a perception that non-ejectors are the poor man's gun and that they were of lower quality, made for keepers and gardeners or for being bashed around

Below: Waiting for bolting rabbits. 19th century shooters did not need sidelock ejectors for this and neither do we.

Bore Size

Shotguns are graded in size with terms such as 12-bore, 20-bore etc but many people are unsure what system is in use to denote the size differential.

The system is an old one dating back to the days of the cannon and is based on the number of balls of equal size that may be made from 1lb of pure lead.

A 10-bore, therefore, is a tube that would exactly accommodate a ball, of which ten of equal size could be made from 1lb of pure lead. Clearly if 12 balls were made, each would be smaller and so a 12-bore is smaller than a 10-bore and a 16-bore is smaller than a 12-bore.

The size of the bore can be measured and this measurement will often be found stamped on the barrels by the proof house instead of, or as well as, the nominal bore size. The following chart shows the diameter in inches of the most commonly encountered bore sizes;

8-bore	.835"
10-bore	.775"
12-bore	.729"
16-bore	.662"
20-bore	.615"
28-bore	.550"

Note that a barrel is considered 'out of proof' if it is ten thousandths of an inch larger than the mark stamped on the barrel at proof. The measurement is taken nine inches from the breech (i.e. a gun stamped .729" (or 12) is out of proof when it measures .739"). It is important to note that there is scope within the nominal bore size for variation – a 12-bore is as likely to be stamped .740 as .729, so reading the original proof dimensions is imperative when measuring the current dimensions to check for wear or repair. Do not just assume it is standard.

Indian swamps or the African bush. There is truth in this perception, but only part of the truth. Non-ejectors were made for reasons other than pure economy.

• One reason was as 'live pigeon' guns. These guns could be very expensive and very well made indeed. Live pigeon shooting was a popular sport with a great deal at stake in terms of prizes, betting and the reputation of the makers of the guns used to win. Ejectors were not important and were not usually fitted. These guns make a good modern choice for the discerning pigeon shooter, inland duck shooter and the sporting clays shooter. They will often be heavy (up to 8lbs), straight stocked with high combs and feature strong actions with cross bolts and 2¾" chambers.

• As 'wild-fowling' guns, ejectors are a complication and they can go wrong. In the mud and rain routinely encountered by the wildfowler of a century past, they were seen as an irrelevance and a potential annoyance, should they malfunction on the marsh. This is no less true today. Wildfowling guns were usually plainly finished but in better grades were of sound construction and good materi-

Two of the author's boxlock non-ejectors: an A. Hill & Sons (top) and an Edwinson Green, taking a rest during a day which accounted for 85 pigeons.

Bore Size Marked at Proof (pre-1955)

Some 12-bore barrels (1887–1954) will be found marked $^{12}/_1$ or $^{13}/_1$. These show variations within the nominal bore size.

A barrel stamped 12 will be .729" while one stamped $^{13}/_1$ will be a little tighter (.719") and one stamped $^{12}/_1$ will be a little wider (.740").

From 1875–1887 guns were often marked 12B (showing the bore size as 12-bore, ie .729") and 14M (showing the muzzle dimensions of a choked barrel as 14-bore, ie .693"). The warning 'NOT FOR BALL' is also to be found on the choked barrel. The practice of showing the choke as a muzzle 'bore' size was discontinued in 1887 and instead, the word 'CHOKE' is stamped on the barrel flats.

This Lincoln Jeffries wildfowling boxlock non-ejector has interchangeable 10-bore and 12-bore barrels and, like many guns of the type, utilizes the Greener cross-bolt.

als. Do not let an apparently austere appearance and lack of ejectors mislead you into the false assumption that it must therefore be a low quality gun.

Boxlocks have good water-resisting properties and are robust when made heavy (as wildfowling guns firing enhanced payloads need to be). For these reasons boxlock non-ejector wildfowling guns are frequently encountered and warrant consideration. They are not to be confused with low quality guns of the same basic format. Greener's 1888 catalogue again provides a useful historical reference; his 'Facile Princeps' gun could be finished and engraved to the highest 'Royal' standard for $282 or the same action described as: 'Plainly finished, well made, Hammerless Gun, without engraving, English Damascus or Siemens steel Barrels' could be had for a mere $82.

Better quality plain guns can be identified by the details: wood-to-metal fit, balance, tightness, proportion, pin alignment and precision, the quality of any engraving present, though it may be minimal, and many other small indicators.

• As keeper's guns. These guns were made for the low end of the market and were utilitarian in nature. However, some very good firms made guns in a variety of grades and those such as Westley Richards, Bland and Greener produced basic non-ejector guns of sound design and good workmanship, albeit without any notable finish or embellishment. Henry Sharp commented in 1909 'A

gunmaker of acknowledged skill and repute in designing best guns and rifles, if entrusted with orders for medium grade colonial or keepers' guns, may be relied upon to give better value for money than the small maker, dealer and jobber can afford to give'. They have an honest charm of their own. Compare a Westley Richards 'keeper's gun with a boxlock AYA 'Yeoman' or a Lincoln over-and-under and see which is the more beguiling.

Greener's 1888 utility gun was marketed as 'The Dominion Gun' (I have found no record of disputes with Holland & Holland over the use of this term, which is surprising as Holland's used this name from the 1930s for their plain Birmingham-made back action gun. Research suggests that Greener no longer used the term by that time.). The catalogue entry describes it as: *'Top-lever, double bolted, snap breech action with extended rib, laminated or Siemens steel barrels, either choked on W.W. Greener's world-renowned method or improved cylinder, to order. In this gun all the value is put into the barrels, locks and shooting'.* It cost $46. Generally, even these cheaper offerings from well-known makers would be strong and serviceable – makers did not want their expensive work devalued by the possibility of lower-grade guns with their name on the locks failing or shooting loose.

'Single barrel guns are worth no consideration'

This was certainly true a few years ago. I would pick up nice single hammer guns from the 1880s and '90s for $140 and nobody else was interested. The single is gaining in desirability however; nothing stays cheap for long and once the anomaly appears, people will move in.

It is still true that a cheap way into collecting and using old guns is the single barrel option. This style of gun is still available for much less money than a double. The exceptions being the 'made to order' singles of the top makers – as boy's guns or trap guns. They are rare, top quality and expensive.

If you have under $200 to spend and want an interesting old gun, do not discount the single. They are very shootable, attractive pieces and are a pleasure to use for rabbiting or rough shooting. I have a nice Hughes 16-bore hammer gun, with skelp barrels and a Jones under-lever, which is my preferred hedgerow gun for spring evenings after rabbits or the odd pigeon.

Pleasure is also to be had with some of the cheap English utility guns of bolt or Martini action, the Webley and Scott .410 or 9mm and the Greener GP being good

examples of the former and latter actions. They are easy and fun to take to bits, re-finish and put back together and are robust and cheap enough to do so without worrying.

Rook rifles are worthy of further mention here – in the .410 smoothbore conversion. Most such rifles were made in large quantity in Birmingham by firms such as W & C Scott and retailed by many well-known makers in London and elsewhere. They come in a variety of grades but some are extremely well finished and are delightful to own, though prices have been rising for some time.

Top: Webley bolt-action single barrel guns were made in .410 and 9mm. These excellent basic guns are still going strong and offer excellent value at auction.

Bottom: They are great first guns for boys like Max.

'Modern guns have single triggers but vintage guns all have the old fashioned double triggers. This must be a handicap'.

'Pointless, troublesome creations that should be consigned to the dustbin.' – this is by no means a universally shared opinion, but it is the one I hold on the matter of single triggers on pre-WW1 shotguns (excepting the Holland & Boss systems, which have proved their reliability when cared for). I suppose that having made such an unequivocal and uncharitable statement about a relatively common feature, I should explain it.

The sporting shotgun of either side-by-side or over-and-under form is basically a functional and durable device, perfectly suited to the repeated firing of tens, indeed hundreds, of thousands of cartridges. As in so many fields, simplicity is desirable in gun mechanics. This is evident from the lock work in enduring designs. Perhaps the Anson & Deeley boxlock is the best example of this, with only four moving parts. There is little to go wrong in the locks of a well-made gun, the jointing should be sound and the thing itself perfectly capable of providing years of good service.

It is unusual to have a vintage gun malfunction during a day's shooting. If it does it will probably be for one of the following reasons: a broken spring, irregular ejecting or some fault with the single trigger. Springs are a fact of life. They last for years and then one day 'ping' and you need to have one replaced – one reason I always take two guns to a shoot.

Ejectors are necessary for driven shooting (arguably) and will need regulating from time to time. Early ejector systems can be problematic and should be carefully checked at the time of purchase, but this is for discussion later. Single triggers are a different matter entirely.

What are single triggers for?

Apparently it is beyond some people to move their finger from the first trigger to the second in order to fire both barrels. Why should this be? Well one reason may be that that the shooter is wearing gloves, another may be that he has very thick fingers. Should you fall into the latter category, try an 'articulated' front trigger; essentially this is a reverse hinged trigger that ensures fingers are not jammed between the two triggers or injured by recoil when firing

the second, back trigger. If you are in the former category, remove the glove from your rear hand or the material from the trigger finger of your glove.

Most people are able to fire two barrels just as fast with a double trigger gun as with a single trigger one, when using a gun in the field. I have had strong arguments put to me that the single trigger is helpful for some sporting clay targets, but remain to be convinced that it really gives any discernable advantage to the sportsman.

The single trigger gun can be provided with a barrel selector to enable you to choose which barrel fires first, (Westley Richards offered these in 1887) but this is time-

This Edwinson Green 12-bore o/u is unfortunately encumbered with an unreliable single trigger, which has proved impossible to regulate effectively.

consuming to do, is not universal on vintage guns – and is easy to forget. Double triggers give instant selection of the right or left barrel, allowing the shooter to use the greater or lesser choke, as appropriate. Double triggers hardly ever go wrong. Single triggers go wrong with depressing frequency.

When buying vintage guns, a wide variety of designs will become apparent with regard to lock type, barrel length and material, forend fastener, stock shape and a plethora of other permutations that will mostly all work and will essentially be a matter of personal taste in the evaluation. Hand-made single triggers frequently do not work.

When gunsmiths were trying to devise a reliable single trigger, they came up with many designs and most of these have their share of problems. Essentially they are unreliable and who wants a gun on which one cannot rely? A friend of mine once spent $1,600 having his single trigger regulated and when he next took it to a clay shoot it malfunctioned. It still does not work properly. Older English single-triggers are a potential minefield. This is not a matter of maintenance; it is a fundamental design fault in most cases.

Good quality modern guns generally have reliable single triggers these days. Your Beretta or Browning is unlikely to go wrong. Mike Yardley pointed out to me that a reliable single trigger mechanism lends itself to machining and modern over-under trigger-plate guns have the system set in metal rather than wood, which is a better arrangement. This reliability claim cannot be made for many of the expensive Italian makers unfortunately and I have some disturbing anecdotal evidence of these guns firing both barrels at once or refusing to fire at all.

When viewing vintage guns with single triggers, it can be difficult to find out whether the mechanism is in good order or not. Many only work when a live round is fired. With snap caps for example you will find a Boss gun needs three pulls to fire both barrels. All are complex, involve fine tolerances and small moving parts: Burrard noted in 1931 ' *Considerably over a hundred different single trigger patents have been taken out; the great majority of these patents seem to be in use; and no two seem to be alike!* ' Lots of potential trouble.

My advice when purchasing vintage guns is to avoid single triggers – it is advice that will help you avoid frustration and expense. Besides, they really offer no advantage over double triggers for shooting game. 'KISS' (*Keep It Simple, Stupid*) applies absolutely when appraising trigger mechanisms.

'Nobody shoots with a side-by-side anymore but I never see British over-and-unders'

I have to admit to a personal bias for British side-by-side shotguns. I would probably shoot better with an over-and-under; in fact Mike Yardley's view is that an over-and-under will consistently outperform a side-by-side by 5%, all other factors being equal. This means I should put more in the bag, break more clays and be a better, more accom-

plished shot with an over-and-under. However, I accept the handicap willingly because I choose to. Shooting should be about more than hitting the target; paradoxical as that statement may sound. Part of my rationale is my own attitude to clay shooting – I can't take it entirely seriously as a sport. I shoot clays quite often, and occasionally enter competitions, but in the back of my mind I believe the true purpose of clay shooting is to make me a better game shot where it really matters: in the field. I therefore use only my game guns to shoot clays and shoot gun down almost all the time, regardless of the rules. I personally find my left eye drawn to the side of the over-and-under, which messes with my eye dominance. Perhaps it's just years of shooting side-by-sides exclusively, but over-and-unders are not for me.

I know that a dedicated band of fellow eccentrics share my appreciation of old British side-by-sides but the modern preference is unwaveringly for the over-and-under barrel configuration. Certainly Macdonald Hastings' assertion in 1981 that *'O/U's have never really caught on in Britain where shooting men feel that the conventional gun handles more sweetly'* is far from a reflection of today's shooting scene in the UK. A recent personal check of the fourteen guns present on a walked up-partridge shoot in Hertfordshire saw 12 using over-and-unders of non-British

The short-lived Purdey 'sextuple-grip' over-and-under. This example sold in 2005 at auction for under $20,000. It is actually a very nice gun to handle and of excellent quality.

make and two using side-by-sides (I was one). So, with this confirmed preference for the stack-barrel, why do we not see many British over-and-under guns on shoot days?

The answer is not that they are unavailable, but that most of us can't afford them. Go to the grouse moor or the 500-bird pheasant day organized by a top sporting agent and you will see plenty of these guns in action.

The great makers all made over-and-under guns in the early years of the last century and they still do. There is the Boss action of 1909, arguably the most elegant, the current Purdey (developed from the original 1913 James Woodward gun and adopted by Purdey in the 1948 takeover of Woodward's) and the trigger-plate action Holland & Holland 'Sporting', the most recent and 'modern' design, which offers the H&H buyer a cheaper alternative to the outstanding sidelock 'Royal' over-and-under, which Mike Yardley called '*the best gun in the World*' when he tested it for *Sporting Shooter* magazine in 2005.

In addition to these 'old firms' David McKay Brown of Glasgow now produces some beautiful over-and-under trigger-plate action guns. There are also new over-and-under guns exhibiting taste and style offered by the resurrected EJ Churchill, Holloway & Naughton and Charles Hellis firms.

As ever, price is the factor that constrains most of us. The Churchill will cost you a few cents under $39,500, the Purdey a little over $100,790 plus VAT, the others somewhere in between. For most of us new may not be an option but used guns offer some interesting, unusual alternatives.

The lowest estimate I have seen in recent years for a popular over-and-under by a top British maker at auction was $31,600 for a Purdey and the hammer price significantly exceeded this. Decent Purdey or Holland & Holland side-by-sides start from around $11,900, so the change to over-and-under involves a steep rise in price.

Rather than give up and resort to a Beretta like everyone else, here is a proposition for the confirmed over-and-under user: look to the less fashionable makers who were producing designs for over-and-under guns in the first decade of the 20th century.

Transitional guns – cheaper and interesting

There are some unusual examples out there. The quest to make the best over-and-under was one of the last great challenges to the gunmaker of 100 years ago and they were all having a go.

There is Purdey's 'sextuple grip' over-and-under, reflecting Athol Purdey's obsession with a strong action but coming at the expense of weight. These are not desirable to most and a good example went unsold at a London auction not long before the time of writing. Another made a modest $9,000 at Holt's in 2005. After WW2, Purdey switched to the more streamlined Woodward over-and-under design and has used this ever since.

Westley Richards made a detachable-lock boxlock called the 'Ovundo', which is less than elegant in photographs but surprisingly pleasant when handled and can still attract reasonable money. Lang, Beesley and Lancaster also made over-and-under guns of their own design. The Beesley is characteristically innovative, one lock being 'upside-down' to ensure the best angle for the tumbler to hit the striker on the bottom barrel.

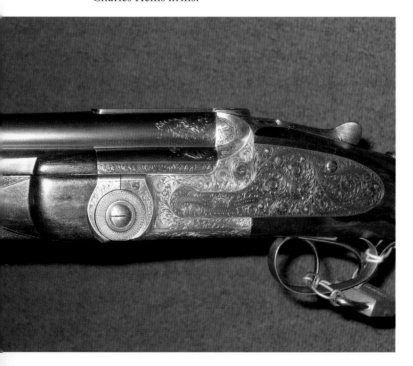

Beesley's 'Shotover' patent action for the over-and-under. The lock firing the 'under' barrel is upside down.

The bargain basement of British over-and-unders

My friend Peter Jones has an interesting over-and-under by Edwinson Green. Green was a prolific inventor and skilled gunmaker based in Cheltenham & Gloucester. His over-and-under design of 1912 was one of the first to show real

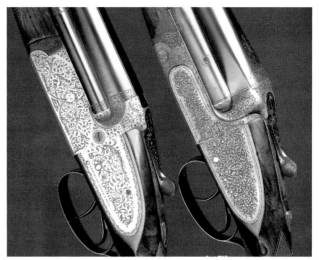

The Edwinson Green over-and-under. The gun on the left is the first ever made to this design.

This Lancaster over-and-under demonstrates the under-lumps that give such early designs their deep actions. This is one of a pair sold in a London auction for $9,180 in 2005.

both guns was nicely executed but very different in style, one traditional scrolling and the other foliate in nature with various nymphs blowing horns. As Peter says, *'Turn up to a shoot with one of these and everyone knows you have something a bit special'.*

These two well-preserved examples of top quality English gunmaking in the preferred modern style of barrel configuration sold for under $7,350 each. (At the time of

potential. Purdey used it as the model for their first over-and-under before switching to the Woodward design.

Peter's example is ribless and light and handles very well. The Green design has been described as 'clumsy' and too tall in the action. This is because it has a conventional lump at the bottom, rather than the now-common, Robertson inspired, 'bifurcated' lumps on the sides of the barrel, as used on the Boss and Woodward guns, but in practice it balances nicely and feels the quality gun that it is. Green guns are usually finished to a very high standard. The workmanship is excellent and the engraving style can be varied and idiosyncratic.

Two notable examples of Green over-and-under guns appeared in the March 2004 auction at Holt's.

One was the very first made to this design, and as such was an eminently collectable item as well as a very shootable weapon. Unsurprisingly it was secured by a trade buyer and was displayed for sale at the subsequent summer Game Fairs at a considerable mark up. The engraving on

Detail of the bites on a 1920 Edwinson Green over-and-under 12-bore.

What detracts from the value of a gun?

The Stock

Cracks. Cracks in the stock are at their most problematic in the 'hand'. This flexes when the gun is fired and thus, any crack is likely to get worse and is hard to fix. Re-stocking is very expensive.

Repairs. Repairs to damaged stocks are rarely satisfactory. They will ward off potential buyers.

Plain Wood. Buyers will pay a premium for good figure in the wood. It is also desirable for the stock and fore end wood to match.

Lengthening Pieces. Long wooden extensions are unsightly and the gun never looks 'right'. Short extensions of rubber 'Silvers' pad types and leather-covered pads are better.

Cheek Pieces. Some guns will be encountered with leather-padded cushions to prevent bruising of the cheek. They may have suited the original owner but will not suit the gun.

The Barrels

Dents. Dents can be raised but the barrels will lose metal each time and therefore weaken. It is a process many buyers cannot be bothered to contemplate.

Bulges. These can require re-barrelling: prohibitive in the case of most medium grade guns at today's prices.

Thin Walls. Guns with thin walls damage more easily but are worth considering if at bargain prices and not intended for very heavy use.

Pits. Pits are generally caused by old corrosive primers using fulminate of mercury to fire the charge of powder. If not thoroughly cleaned, the barrels would start to rust and the rust spots then increased in size until becoming visible as deep holes in the bores. Pits can be lapped out at the risk of over-enlarging the bore. A few shallow pits can be lived with, just clean the gun thoroughly and oil the bores well before storage (store muzzle down).

Enlarged Bores. Where pitting has been removed, the bores will be wider than originally proofed. If more than ten thousandths of an inch bigger, they are 'out of proof'.

Sleeving. Sleeving extends the life of a gun and is much cheaper than new barrels. However, a sleeved gun will be worth considerably less than one that has not been sleeved.

New Barrels. A gun with the original barrels in good order will be worth more than the same gun with new barrels. If the new barrels are by anyone other than the maker of the gun, the price will be lower still. Consider this – are you buying a gun for hard use? If so a bargain with replacement barrels could be exactly what you are looking for.

Shortening. Barrels reduced in length will be damaged as to their shooting quality, balance and originality and are therefore to be avoided unless cheap and intended for customization.

The Action

Unfashionable Actions. Trigger-plate action, lever-cocking Grant guns are cheap, back action sidelocks of the transitional period likewise. Anything odd-looking will fetch less than something resembling current production. These guns are very usable and can be excellent value and fun in the field.

Operating Levers. Top-levers & side-levers are desired above all others. Prices are affected accordingly.

Engraving. Very fine engraving will add to the price. A Webley Model 701 boxlock recently made $4,000 at auction. The gun is the same as the Model 700 except better engraved and with drop points on the stock. A Model 700 will make between $700 and $2,000.

Other

Faulty design or conversion. Single triggers that do not work, ejectors that do not work, ejector systems that are later conversions and not to the preferred design later adopted by the maker.

Lack of a 'famous name'. A famous name will make a gun more expensive. A Scott 1882 back-action sidelock with 'W&C Scott' engraved on the locks will fetch less than the same gun with 'Holland & Holland' on the lock plates despite the fact that Scott made the Holland & Holland guns. As Donald Dallas tells us in *Holland & Holland: The 'Royal' Gunmaker* '…before the establishment of Holland's factory in 1893, the firm were not producing the vast majority of their weapons.'

Dallas continues, 'Most of this evidence points to the Birmingham firm of W & C Scott as the supplier of Holland's products.' Scott's and other Birmingham firms were exceptionally skilled gunmakers yet the public perception of 'London' guns being better products persists and is visible in pricing. Buying according to the quality evident in the gun rather than being beguiled by the name is a policy that will save you a lot of money and enable you to consider some fabulous guns by 'lesser' makers.

John Farrugia of the Cheshire Gun Room measuring wall thickness at auction. This is essential before buying.

writing the lowest priced British over-and-under is the E.J. Churchill 'Imperial' trigger plate gun at $39,425). This shows that the practical eccentric can get hold of a gun with bags of style and an idiosyncratic nature for money similar to that required to purchase a decent quality foreign over-and-under.

An unusual customizing option

As a brief aside, casting a glance to the user of the Beretta who may not yet be ready to embrace the vintage English gun but is beginning to be tempted by something a little more individual, subtle and elegant, let's consider an unusual offering from a current English gunmaker.

For the, largely practical, novice eccentric, A.A. Brown will take a normal Beretta and customise it to make it resemble, and handle more like, a traditional British over-and-under.

This involves a total reworking of the stock, forend and triggers and much re-filing of the furniture. The results are very pleasing.

Two-inch guns and other light-weight influences

Most modern guns are designed to accept cartridges of 2¾". The traditional English game gun, especially of pre-WW2 vintage is usually chambered for the 2½" cartridge but there is another, little known, cartridge that enjoyed a brief period of popularity early in the 20th century and the modern 20-bore aficionado may find this an interesting vintage proposition. The item in question is the two-inch 12-bore cartridge.

The origins of the 2" chamber

The 2" chambered gun was developed in response to the quest for lightweight game guns. There were numerous attempts in the late 19th and early 20th century to develop a gun that had fast handling, was lightweight with low recoil but without losing the quality of ballistic efficiency or 'hard hitting' as old gun books usually put it.

Attempts by Charles Lancaster to introduce his, fully loaded but measuring a little over 2", 'Pygmy' cartridge failed amid bad publicity on the tendency of the shot to ball when used in some guns not specifically chambered for it. Teasdale-Buckell is less than complimentary about the Pygmy in 1900's *Experts on Guns and Shooting*.

However, Lancaster's failure (generally attributed to the shortened wad having a tendency to tip, allowing hot gasses to escape and fuse the shot) did not kill off the idea of shorter cartridges for use with nitro-powders and the idea became seriously fashionable in the 1930s. The shooting man wanting a light gun, could move to a smaller bore gun or try a lightweight 12-bore. From 1924 Lancaster marketed (and some other firms subsequently also retailed)

Guns with 2-inch chambers are invariably marked like this for obvious safety reasons.

the 'Twelve Twenty', described earlier, designed to have the handling characteristics and dimensions of a 20-bore but chambered for a standard 12-bore cartridge. Lang applied some of this thinking to his bizarre *Vena Contracta*, essentially a 20-bore with 12-bore chambers.

Thomas Turner of Birmingham & London had earlier made his *featherweight* hammer gun in twelve-bore weighing in at around 5 ½lbs. This has a tiny forend and much of the wood removed from the stock to save weight. Lincoln Jefferies of Birmingham also specialised in lightweights. The gunmakers' continuing quest for the ideal lightweight gun led to the further development of the short cartridge, earlier tried by Lancaster.

Safety was clearly a consideration in the thinking behind developing a lightweight cartridge for the 12-bore. Many shooting men (for good reason) do not like the idea of 20-bore cartridges and 12-bore cartridges getting mixed up – always a possibility if you own both. The twenty-bore cartridge can lodge in the forcing cone of the twelve-bore and allow the twelve-bore shell to sit on top. Upon firing, the barrel will burst. The adoption of a 2" twelve-bore means this can never happen, as the sportsman need only ever have twelve-bore shells. Among those who offered 12-bore

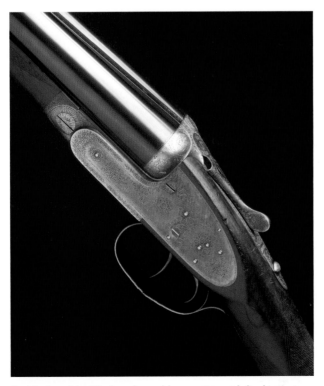

A side-by-side 2" chambered bar-action sidelock ejector, made by Purdey in 1935.

guns chambered for the 2" cartridge were Purdey, Holland & Holland, Westley Richards, Rosson, Linsley, Graham, MacPherson, Stephen Grant and Skimin & Wood. The guns were built to weigh less than 6lbs and intended to fire ⅞oz of shot or less.

Ballistically speaking, when firing any given load in two guns of different bore, the larger bore will generally give the more even pattern. The traditional ideal was the 'square load' with the lead pellet column being equal in height and breadth. Pellets stacked in long, narrow configurations lead to longer shot strings and more deformed pellets. Therefore a lightly loaded twelve-bore is generally considered better than a heavily loaded 20-bore. The lightly loaded 2" cartridge in a light 12-bore should pattern very well and deliver its shot in a nice even pattern with a shorter shot string and fewer deformed pellets than a similarly loaded 20-bore.

Drawbacks

Major Burrard, writing in the 1930s, was sceptical about the ballistic efficiency of the 2" 12-bore, believing it at the extreme of the above principle due to the light shot load providing insufficient resistance to the burning of the powder. Others disagreed and the case is inconclusive but the market remained unconvinced of its benefits and the use of the 2" cartridge faded, most makers building 2½" chambered guns down to 6lbs for use with 1oz loads.

The trend for ultra lightweight guns became absorbed into mainstream gunmaking. The average weight of a twelve-bore game gun had been reduced from around 7lbs in the 1860s to nearer 6lbs by 1910. Quality game guns in 12-bore weighing between 6lb 1oz and 6lb 3oz will be encountered regularly – though the lighter weight comes at the expense of durability – many such guns have very thin barrel walls. Another factor is that a lighter gun will not have the same capacity to handle recoil as a heavier one. I personally prefer a 12-bore to weigh in at around 6lb 10oz -but I am 37 and in good health. I may feel differently about it when I am 60.

Revival

Despite the modern norm for guns to be chambered for the 2¾" cartridge rather than the old standard 2½", the 2" is still a very popular chamber size in America for shooting quail and walked-up game and the demand for vintage guns of this type at auction has risen sharply in recent years. The cartridges are still available in the UK from major manufacturers, making it a usable gun in every way (try www.justcartridges.com for a good choice).

Damascus versus steel barrels

Modern gun barrels are made from steel. The technology exists to make solid steel ingots, form them into long bars and bore holes in them to make tubes. Since around 1900 this has been almost universal practice in the manufacture of shotgun barrels.

This has not always been the case. In the very early days of gun barrel production, a sheet of metal was bent around a mandrel and the seam hammer-welded

Greener's '96-1 Rule'

Greener's rationale was that there was no point building a gun so light that it could not absorb the recoil delivered when a given shot load was fired. The rule may be a little out of date with the varieties of modern cartridges on offer and tinkering with different loads and high or low velocity shells may produce a combination that works outside Greener's principle but the fact remains that a gun that is too light for what you want to fire through it, is a handicap. Greener last published this in his 1909 edition of *The Gun and its Development*.

Load	Gun
1 oz of shot	6lb gun
1⅛oz of shot	6¾lb gun
1¼oz of shot	7½lb gun

closed. This left a long, straight weak spot and made the barrel liable to burst under pressure. Greener called them *'worthless tubes'*. The stresses put upon a barrel are upon the entire inner surface area as the charge expands outwards and a pressure wave is pushed along the tube. A seam that runs around the circumference withstands this pressure better and the practice of wrapping the iron and steel in long ribbons around the mandrel and hammer welding the

tubes as they were formed became the norm from around the first decade of the nineteenth century.

This type of barrel, known generically as 'Damascus', with the intricate swirling patterns made by the mingling of the different grades of iron and steel, is to many eyes the most beautiful type of barrel ever manufactured. It is now a lost art. The actual patterns come from the edges of the different ribbons of iron and steel being carefully positioned on the outside while the red-hot metal was hammered into shape. The intricacy of the pattern is dependent upon the number of rods or ribbons used. Best quality English Damascus, largely produced in the Midlands, was generally of two or three rods or 'irons'. Belgian Damascus was often made using more rods and produced very finely figured barrels.

Greener famously wrote, 'The English maker will take a barrel that will do best; the foreign maker, the barrel that will look best'. Proof house trials of available Damascus barrels demonstrated that English 'two-iron' and 'three-iron' Damascus was of equal strength to the new steel barrels, Greener believed best Damascus to be better than Whitworth steel. Many British makers used tubes of both English and Belgian manufacture in all grades of gun, depending on prices and availability. Writing in 1891

Gun-barrel iron, twisted, and laid into a riband.

in *Modern Shotguns*, Greener was scathing of the practice: 'the greatest blame attaches to those London gunmakers who have long used and continue to use an inferior Belgian imitation of a genuine English product, and which they know , and have long known, to be inferior'.

Damascus was nevertheless gradually replaced by steel as the preferred barrel material of the Gun Trade once the 'fluid pressing' system of manufacture was perfected in the 1880s. This was a process that involved putting the sides of the liquid steel ingot under pressure as it cooled, removing the problem of air holes forming in the middle as it set. Forged steel of high quality was, at last, available

Importing and Exporting your Vintage Guns

The UK is a great source of guns for the collector and enthusiast. Many dealers specialize in buying guns in the UK gun trade or at auction and exporting them for sale in the USA or elsewhere. Such dealers obviously do this for profit and make a good living from doing it. If you want to play their game and buy in the UK, you need to make sure you are not stuck with a gun you can't get out of the country. First, you need permission to buy it. British subjects need a Shotgun Certificate but visitors to the country can apply for 'Visitor's Shotgun Permit' from the police.

The Visitor's Shotgun Permit is valid for up to 12 months and is valid throughout GB, regardless of the police area that issued it. The permit will enable the visitor to go to a gun shop or auction and buy a shotgun, as long as it stipulates that the holder has permission to buy as well as possess shotguns. Sometimes a permit will be issued to buy a particular gun, sometimes permits will be less restrictive but the whole process can be troublesome. Once you have your gun – you need to get it home.

To export a shotgun requires a licence from the Department of Trade and Industry:

Department of Trade and Industry
Export Licensing Unit
4 Abbey Orchard Street
London SW1P 2HT
Tel: 0207 215 8070
E-mail: eco.help@dti.gsi.gov.uk

If worried about the time, effort and expense involved in arranging the export of a vintage firearm, there are established firms who will help or even arrange the whole thing for you, for a reasonable fee. Auction houses will also pass the export arrangements on to such people. The Cheshire Gun Room are experienced practitioners and can be contacted for advice:

John Farrugia (The Cheshire Gun Room)
29 Buxton Rd, Heaviley,
Stockport, SK2 6LS, U.K.
Phone: +44 161 480 8222
Fax: +44 161 612 6602
Mobile: +44 7973 333129
info@cheshiregunroom.com

Age of guns

US law does not recognize guns made before 1898 as firearms and they are not subject to restriction or special procedures for import into the country. Therefore you can legally take your 1875 Westley Richards boxlock from London to Dallas without a US import licence – but you still need a UK export licence.

Sending guns to the UK for repair

The procedures involved in getting a gun to the UK from the US are simple, just put it in the post (or DHL) and send it to a UK registered firearms dealer. He will repair the gun and then the fun begins – you need a UK export licence and a US import licence to get it back again.

While researching this book, one American contributor told me of his problems sending a gun to be proved in the UK without any problem but the proof house would not send it back to him because he did not have an export licence.

The solution

The easy way to do this is to go to a Federal Dealer in the USA, who will arrange the import licence necessary to get your gun back after the work has been done. He will deal with a colleague in the UK who will arrange the export licence to get the gun returned and arrange for all the work to be carried out and deal with logistics for you. You just take your gun to your Federal Agent and collect it from him when all is done.

Fees will be around $100 for processing to the Federal Agent, $90 for processing to the UK agent and the expenses of packaging, transport etc, at cost. Unit costs are considerably reduced when sending more than one gun. John Farrugia at the Cheshire Gun Room will arrange everything necessary and can provide a current list of approved Federal Agents in the United States.

He will also arrange for shipment of guns purchased in the UK at auction or privately and can arrange a similar service to Australia and New Zealand.

Doing it Yourself

Occasional Importer Procedures For U.S. Individuals (For personal sporting use – not for re-sale) by Mick Shepherd

Use Form 6 obtainable from U.S. Bureau of Alcohol, Tobacco and Firearms.

Telephone the ATF Distribution office 703-455-7801. Have them send you 3 copies of *Form 6* (you can put more than one gun on a form). Alternate number: call ATF Information 202-927-7777.

As an 'Occasional Importer' include following information on Form 6: 'Application to Import' Maker, serial number, gauge, length of barrels and total length of the gun.

Any FFL holder near you may fill out the form. He need not hold an importers licence. YOU are the importer and by law may do so as an 'occasional importer.' U.S. law requires this be for personal sporting use; indicate this on your application in the appropriate space. All else is self-explanatory.

When completed and signed by your FFL holder, mail or Fed Ex the form to the ATF. Fed Ex or UPS expedites delivery.

F & E Imports Branch, BATF, 650 Mass Ave, NW, Washington, DC 20226

It may take 6 to 8 weeks for ATF to approve and return your application; it will be stamped NOT FOR RESALE.

Keep the original approved form. Make photocopies to mail or Fax to the vendor. We suggest you have the guns returned by international mail from England to you. This way they enter the U.S. Postal system, go to a customs facility where they are cleared and then directly to your FFL by mail. Duty, if any, may be paid to the postmaster or mailman. Avoid Air Freight.

Your approved Form 6 copies will be inside the package and in an envelope marked ATTN: U.S. CUSTOMS.

Antique Guns (made before 1898) are not considered firearms in the US but they are in the UK.

to barrel makers and it could be bored out to make regular tubes perfectly suited to shotgun manufacture.

The demise in Damascus can partly be explained by the growth in popularity of choke boring after the 1875 'Field Trials' and instances of inferior Belgian Damascus failing or bulging under the added pressure but the main reason was probably cost. Damascus tube production was costly in the man-hours of skilled workers and in materials; the preparation of iron and steel and the hammering into barrels involved significant loss of material at each of the many stages involved. Eighteen pounds of iron and steel were needed to produce a pair of Damascus barrels that, when finished, would weigh no more than 3½lbs, according to Greener (who had his own barrel-making facilities in Birmingham). This simply became outdated as a mode of production.

Charles Lancaster, in the 9th edition of *The Art of Shooting*, wrote '*Damascus' or 'gun-barrel iron' was in use for many generations and the beautiful figure and color contrasts of old barrels are still a delight to the eye. That bursts were not frequent is a great tribute to the skill and workmanship of the old barrel filers in producing such a heterogeneous material. But the primitive and tedious methods of production involved could not survive against the advances, particularly since 1914, in our knowledge of steel, which is the same all through. To the sorrow of many, the question of 'Damascus or steel' has been settled by the fact that the former is not now made in this country, and is probably not in commercial production, in good quality, anywhere in the world'. (1937)*

Another worry for the gunmaker, that steel barrels alleviated, was the possibility of 'greys' emerging in a pair of high quality Damascus tubes. Greys are small bits of waste matter, made during the hammering, that will not color like the iron and steel when 'browned'. Sometimes they do not emerge until the barrels are finished, or even later and customers were likely to complain should they do so. In point of fact, greys do not materially affect the strength of the tubes.

Webley's 1882 report on Belgian Damascus forging, published in Walsh's *The Modern Sportsman's Gun and Rifle* lamented the state of barrel production in the UK thus: '*for many years we have been almost entirely dependent on one maker* (Marshall of Birmingham) *for Damascus, stub Damascus and laminated steel iron; he having a monopoly, has not cared to trouble himself to keep his iron up to its original good standard, notwithstanding the fact that, in consequence of its high price and want of clearness, his trade has been gradually leaving him and going to Belgium'.*

Certainly, Purdey were using Belgian tubes at this time. The main attraction was their freeness from 'greys', though James Purdey the Younger is reported as saying that English Damascus of the best quality was superior, thereby concurring with Greener. The reason for the lower number of 'greys' in Belgian tubes is often attributed to the smaller, cleaner-burning furnaces.

However, despite his complaints about English Damascus, Webley's investigation into Belgian Damascus production echoed Greener in its conclusion *'they* [the Liege barrel makers] *only studied three things: First, to get the greatest possible distinction in color (black and white when browned) between their iron and steel; secondly, regularity of figure; thirdly, clearness of iron; and that whether they* [the barrels] *were hard or soft was of no consequence to them'*

The most famous brand you will find stamped on barrels from the transitional period, (1880s) when steel began to replace Damascus, is *'Made from Sir Joseph Whitworth's fluid-compressed steel'*, along with a distinctive 'wheat sheaf' logo, although *'Krupp', 'Siemens'* and *'Vickers'* are also names you will find stamped on steel barrels. In 1884, Whitworth steel tubes cost a massive 90 shillings compared to the 31 shillings asked for best English hand-forged 3-rod Damascus. However, the way ahead was clearly mapped and the days of Damascus were numbered. Greener's cryptic comments in 1888 that *'They (steel tubes) offer certain advantages to gunmakers which it is not the object of this book to disclose'* suggest that steel was becoming economically more viable as part of the production process.

Damascus became regarded as old-fashioned and in some circles it was thought dangerous, especially in the USA, where large numbers of inferior quality barrels were to be found on cheap export guns made in Belgium and often spuriously signed with the names of 'English makers' like *'Wesley Richards'* or *Horace Greener* or *Purdy*. Through these guns failing, Damascus as a material got a bad reputation.

This relic of the barrel-making process demonstrates how three bars were forged into a single tube.

However, there is no real difference in the qualities or performance of Damascus and steel barrels in the highest grades. Quality Damascus is the practical equal of steel and may be used with confidence, even with modern nitro loads, as long as the gun carries nitro proof marks and is in good condition. Damascus does tend to be heavier than steel though.

I have several guns with Damascus barrels and use them regularly for game and sporting clay shooting. Regarding performance, an interesting comment from Lord Walsingham was that Damascus did not 'ring' with the severity of steel barrels and was less likely to produce a 'gun headache'. It was not unheard of in the transitional period for guns to be made with the new steel barrels, only to be returned to have Damascus replacements made. Certainly Purdey, for one, have records of this happening. Lord Walsingham wrote in 1888 after shooting a record 1070 grouse in one day, using four Purdey hammer guns *'I have learned today that Whitworth steel barrels are not desirable for a heavy day's shooting... I am now replacing them with Damascus, as in all my other guns'.*

The marketplace has changed its attitude to Damascus gun barrels in recent years. Not long ago, Damascus would detract from the value of a gun, now it seems to affect it much less and I detect a continuing shift. I would not be surprised if Damascus barrels of top quality and good condition soon begin to fetch a premium. Henry Sharp wrote in 1909 *'Despite the fact that present indications point at the disappearance of the figured iron, it would not be surprising if sportsmen of the future were to return to the early love'.* I think Henry could be right.

Remember – there will be no more Damascus, ever. Nobody has the skills to make true Damascus anymore so a fine pair of Damascus barrels on a gun should be appreciated for the excellence and rarity they represent. As a caveat to the last statement, it has come to my attention that some stocks of Belgian Damascus barrel tubes of high quality have been discovered in storage and made available by Peter Dyson in the UK. There are also rumors of some experiments in Sweden with modern production of Damascus-style tubes, though the process is not 'Damascus' as the old barrel makers would have known it.

Perhaps, like the return of the wild boar and the great bustard to the English countryside, after years of being hunted to extinction, we shall again see guns being produced by English makers featuring Damascus barrels.

Damascus patterns vary according to the number of 'irons' used and color depends on the browning recipe. It can range from 'Colman's mustard', through chocolate-brown to black and silver.

In fact, W.W. Greener, now under the ownership of W.W.'s great grandson Graham Greener, are planning to make some new guns with Damascus barrels, from limited stocks of English Damascus tubes retained from the old Greener barrel-making works, where production ceased in 1905.

Finally, a word of practical advice: when buying a gun with Damascus barrels, remember that it will have 'dovetail lumps', rather than the 'chopper lumps' of best steel barrels. This is no indication of inferior quality (Damascus tubes were made in the chopper-lump style by Parsons of Birmingham in the late 1860s and early 1870s with hard steel lumps welded onto the softer Damascus

This modern version of Damascus steel, produced in Sweden by Damasteel, is being used for new guns, though it is not manufactured the same way as traditional Damascus.

tubes during the forging process) but the join needs to be checked for integrity as part of the overall evaluation of soundness. Early steel barrels were dovetailed in the manner of Damascus, but this method was superseded by the forging of the lumps as integral parts of the tube, in the now commonly used 'chopper lump' method employed on best guns.

Damascus terminology

Prior to the development (in the 1880s) of the technology to produce steel barrel tubes in one piece, barrels had been made by a process of hammer welding different grades of iron and steel around a mandrel. The resulting, figured, barrels are generally referred to as 'Damascus' by modern writers. However, 'Damascus' barrels were made in a wide variety of grades and the manufacturing process varied accordingly. Here are some common terms encountered when reading contemporary writers, about types of Damascus barrel and what they refer to:

Laminated Steel

This was made from best quality steel scrap mixed with some charcoal iron and worked under a forge hammer repeatedly until the close and even grain desired was achieved. The metal was then rolled out and shaped into a tube in the conventional manner. Greener praised the practical, hardwearing qualities of laminated steel above anything else then available; *'Steel barrels, even of the best quality, will not stand heavier charges than the best barrels of English laminated steel'*. Unfortunately, from an aesthetic viewpoint, it lacked the intricate pattern of other forms of Damascus barrel. But then, so does steel.

Stub Damascus

This was usually made by heating old files, quenching the red-hot metal to make it brittle and then pounding it into very small pieces. This was added to a quantity of nail stubs from horse-shoe nails. The mixture was heated in a furnace to fuse the component parts. The metal was then hammered into rods and twisted and welded into tubes in the usual manner.

Scelp

This was a cheap variety of barrel material also referred to as 'Twopenny' or 'Wednesbury Scelp'. Scelp is 'single twist' variety of tube, the twist running from breech to muzzle. The iron used was of lower quality scrap but if well forged and hammered could be serviceable. Scelp tubes were generally made thick and heavy and not used for good quality guns.

Illustrations from Greener's 1909 edition of *The Gun and its Development* showing some English and Belgian tubes then available to gunmakers.

Sham Dam

Sham Dam is the term used when a poor quality steel barrel was formed by welding sheet steel along its length and then wrapping a thin Damascus layer over the top to deceive the buyer (or gunmaker) into believing the barrel to be a higher quality one of Damascus construction. This deception was not uncommon and when done skillfully could fool gunmakers as well as the layman. The barrels were predictably prone to failure.

Pointille Twist

This type of Belgian tube was very well figured and free from 'greys' and became popular with some British gunmakers in the 1880s. It looked attractive but lacked the toughness of British Laminated Steel.

Crolle

This was a predominantly Belgian-made Damascus of fine figure using three or four irons twisted and hammered to give excellent figure. Greener dismissed it as *'a pleasing but otherwise useless deviation from true Damascus figure.'*

English Steel Damascus

This was composed of steel and iron in six-parts-to-four proportions. Tubes made with higher steel content, usually eight-parts to two-and-a-half of iron were termed 'Silver Steel Damascus'. In either case, the best barrels used three or more twisted rods of metal and were well figured and tough.

Comparative strengths

There was an interesting Birmingham proof house experiment to discover the relative strengths of the available barrel tubes carried out in April 1888. The test-barrels, equal in size and wall thickness, were each in turn loaded with a charge of powder and shot and fired. The shot and powder charge was increased until a failure became apparent in the material (a bulge), at which point the result was recorded and the barrel eliminated from the competition. In all, 117 barrels (39 types) were tested, all of .729" with identical external dimensions. The test was a kind of 'last man standing' contest and the results were surprising:

1st English three-rod machine-forged Laminated Steel.
2nd Whitworth Fluid-compressed Steel (by far the most expensive of the three).

Powder Development: Key Improvements

1313 Schwarz reputedly 'discovers' gunpowder

1412 Gunpowder manufacture in England becomes established

1751 Gunpowder in its 'modern' form is standardised (Saltpetre 75%, Charcoal 15%, Sulphur 10%)

1866 Colonel Schultze produces a 'smokeless' powder of nitrated wood mixed with barium or potassium nitrite

1882 Reid introduces E.C, a 'smokeless' gelatinised powder

1888 Nobel introduces 'Ballistite', made from nitro glycerine absorbed in soluble gun-cotton with added camphor

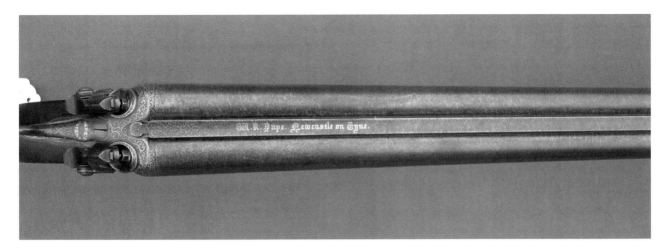

The rib traditionally carries the name and address of the maker. Note that the rib has lifted at the breach end.

3rd English machine-forged two-stripe Damascus.

4th Siemens-Martin process English steel.

5th English hand-forged four-rod Damascus.

Interestingly, foreign (Belgian) Damascus, though popular because of its fine figure, showed poor quality when tested for strength. The best placed Belgian Damascus barrel featured at 18th in the table (Pointille Twist) and the next best was 'Crolle' in three rods and four rods, which were placed jointly at number 25 in the table of merit.

Cheap English 'machine-forged skelp twist' priced at 12s for a pair of barrels featured eleven places higher than the Belgian 'Crolle' four-rod Damascus, which cost 35s; showing that cost, beauty and strength were not necessarily linked.

The report table features machine-forged Damascus, hand-forged Damascus and the new steel tubes competing closely for top honours. Forged steel became better in the years that followed and the industry lost the skilled workers needed to make best Damascus, as industrialization replaced craftsmanship in barrel-making.

The rib

The top of the barrels, where the tubes join, forms a sighting plane and this is usually fitted with a rib of metal soldered (or 'tinned') into place. The British preference for this use of tin and rosin flux instead of brazing is rationalized by the relatively low heat required to perform the operation.

This ensures the barrel steel is not affected structurally during rib-fitting. Tinning is strong enough to withstand hard use but enables a rib to be re-laid easily, should it become necessary.

It has the added benefit that should water get under the rib, the tin will not rust or allow water to penetrate further into the joint between the barrel tubes.

This picture from the Holland & Holland factory shows the rib, wired into place and held with wedges, being laid on a set of barrels.

The shape of the rib, its height and profile, vary for some practical and historical reasons. Whatever the style, the rib should bear close inspection for evenness, straightness and central positioning. You can tell a lot about the quality and condition of a set of barrels by inspecting the rib closely. Some of the more common types encountered are described next:

I often wonder about the practical importance of rib. I have been told by instructors that sunken rib and rib-less guns, for example, are harder to point straight than other types but since one is supposed to look at the bird and not the rib when shooting, theoretically it should make no difference at all.

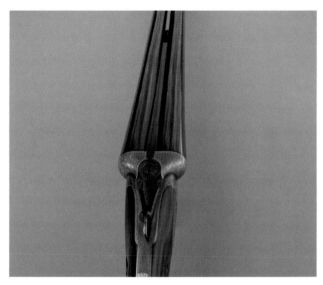

The Rib-less gun. This enables the gun to be made lighter, dispensing with the metal of both top and bottom rib. The barrels are joined at the breech and muzzles and the conventional rib extends part-way but the remainder is not connected. Alex Martin is noted for his production of the 'rib-less gun'. Many were made for Martin by A.A. Brown.

The Concave or 'Game' Rib. This is the most common rib found on game guns and will be the 'default' style of rib unless a special one is ordered. It should taper nicely to the end bead.

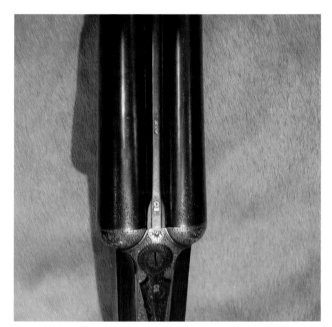

The Churchill Rib. This is an attractive rib, raised and file-cut, but narrower than a normal 'pigeon rib' and tapering towards the muzzle. The idea is to give the impression of length to short barrelled guns. It is most commonly found on Churchill XXV guns and others of like type with 25" or 26" barrels. It is also believed to help move the point of aim higher, to counteract the downward muzzle flip of a short barrelled gun.

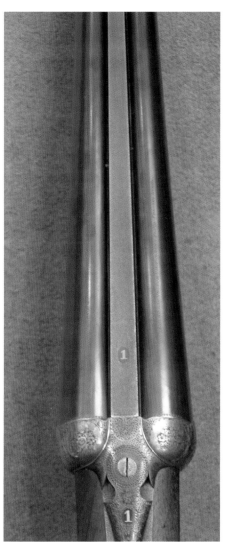

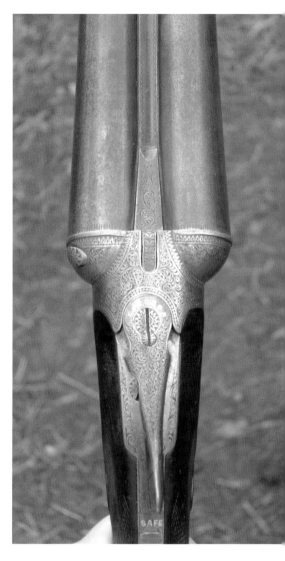

The Sighted Rib

A shotgun barrel typically has no back sight as the shooter's eye effectively acts as one. However, some shoguns were designed especially to shoot solid projectiles as well as conventional shot loads. Sights on such guns are usually of the flip-up leaf variety and may be preceded by a filed, flat rib similar to that on a live pigeon gun. The photograph shows the rear leaf-sights on an 1897 Holland & Holland *Paradox* 12-bore.

The Raised & Filed 'Pigeon' Rib

This is higher, wider and flatter than the game rib and is file-cut on the surface to reduce glare. It will usually be found on guns made for competitive 'live pigeon' shooting . Some wildfowling guns also feature this rib.

The Sunken Rib

The sunken rib usually begins with a ramp-like bridge of metal between the breech face and the rib, where the barrels join. The rest of the rib is below the height of the barrel tubes. Occasionally, guns with sunken ribs are encountered without the bridge from the breech face. These look odd, as part of the breech face is exposed, but they are sometimes encountered even on quality barrels by fine makers.

Points to watch out for in a rib

The laying of the rib can be an indicator of the quality of the gun. If it is not evenly laid or straight, then this is likely to be one of many indicators that the gun is not of high quality. On a gun that otherwise appears well made, rib irregularities could indicate repair work carried out – perhaps the rib came loose and had to be re-laid or it could be an indication that the gun has been sleeved – look for other evidence of this. If the rib has become loose, or if water has been allowed to penetrate the joint between tube and rib, rust can form unseen in the recess, weakening the barrel. If in doubt, pursue your hunch and investigate further. Get expert advice or leave well alone if unable to establish integ-

The rib is a source of potential trouble in old guns. Once rust gets under it, serious unseen damage can occur.

Hump Back Rib. As has been shown, the styles of rib so far described are only representations of almost infinite individual variations on the major themes. The rib pictured above comes from a set of barrels made by William Ford for live-pigeon shooting. The barrels are only 24¼" long and the rib starts level with the fences, rising in the middle and thereafter falling to the bead at the muzzles. Unless the shooter keeps his head in an exact and unvarying position, he loses sight of the bead.

rity. As with so much relating to guns, honest wear is to be less feared than disguised or poorly repaired faults.

When buying British guns, be aware of, but do not be overly put off by, a loose rib. The English gun trade soft-soldered the ribs rather than braze them on, as was the continental practice. This requires less heat and does not affect the barrels' structural integrity. It is a relatively simple process to have ribs re-laid. The layer of tin which fills the space under the rib, in the valley between the tubes, also provides a corrosion-resistant barrier, should water get under the rib. However, any gap should be immediately brought to the notice of a gunsmith before further damage can take hold. I was recently shown a Boss sidelock that had to be sleeved because the (titled) owner had resisted his gunmaker's pleas to get the rib re-laid because there was a hole in the joint, which allowed water access. He continued to use the gun until rust made the barrels unsound. This penny-wise, pound-foolish attitude has devalued his gun by around $14,000. He also had to pay $1,700 for the sleeving, as opposed to the $400 or so that immediate rectification of the problem would have cost him.

The choke

The origins of choke boring are disputed and the minutiae of these debates are not for this book. Suffice to say that it was likely being used in the United States some time before it first appeared in a patent in England. This first patentee was Pape of Newcastle (May 1866), though he did very little to develop the idea.

Whatever its origins, the man to really bring choke boring to the attention of the British shooting public was W.W. Greener. He proved the benefits of the system in *The Field* trials of 1875 and from that point onwards choke has become another of those contentious issues about which shooting people all have a view.

I have a knowledgeable shooting friend who believes full choke in both barrels gets the best results and another, equally experienced and knowledgeable, who prefers true cylinder and three quarter choke. They cannot both be right. Or can they?

So, what is choke and how does it work? Basically, it is a tapering of the barrel's inner dimensions at the

muzzle. It works to concentrate the shot pattern and make the gun more effective at longer ranges. This description is a simplification, as there are various types of boring which provide a similar effect – recessed choke or 'back-boring' for example. However, the user of vintage guns will find choke enters into the equation for the following reasons:

The gun you have or are considering buying:

• has no choke (i.e. both barrels are bored 'cylinder') because it was made before choke was the norm.

• has chokes you believe do not suit you or the purpose for which you intend to use the gun

• has had the chokes removed because the barrels have been shortened (30" barrels are trendy again now but for many years were considered old fashioned. As a result many guns were shortened, usually to 28", but frequently shorter).

There are remedies to all the above problems: it is possible to have chokes eased if they are tighter than you want but I would always recommend patterning the gun first. Gunmakers would regulate a gun until it managed the maximum number of pellets in an even pattern within a 30" circle at a desired range.

For example a customer could ask for the gun to be regulated to shoot an even 30" pattern at 30 yards with the right barrel and the same pattern at 40 yards with the left. This required more work than simply making the required calibration of so many thousandths of an inch at the choked portion of the barrel. In short, measuring with a choke gauge is *at best* a crude indicator of the pattern the gun is likely to

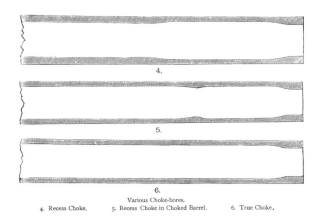

Various Choke-bores.
4. Recess Choke.   5. Recess Choke in Choked Barrel.   6. True Choke.

Choke profiles from barrels illustrated in Greener's *The Gun and its Development*. It is easy to see how shortening barrels has a serious effect on performance.

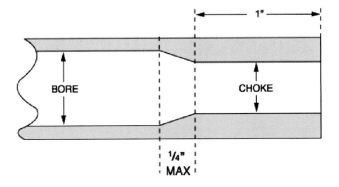

The Purdey fixed choke, showing a short taper and long choked section. This is the choke profile Purdey currently employs.

produce. It is much better to test it before altering it. This type of experiment can be informative and fun.

If guns have been shortened, they are likely to have had the choked portion of the barrel removed. As shown in the diagram on the previous page, the choke portion of the barrel is frequently only around 1¼" in length. Shortening will seriously affect the shooting qualities of the gun and barrels will need attention and regulation to produce reliable patterns. The usual answer was to 'recess choke' the shortened barrels. This involved removing metal from behind the muzzle portion so that the shot moved into the wider section before being constricted again when it met the true bore, just before the muzzle. This can improve the shooting of the gun but it only works in guns with fairly thick barrel walls, it weakens the barrels and is generally only able to manage the equivalent of 'Improved Cylinder' in tightening patterns.

Another option is to add screw-in choke tubes of the type found in many modern over-and-under guns. Teague Engineering will do a very neat (outwardly invisible) job for you, again provided there is sufficient thickness in the muzzle walls.

However, there is always doubt regarding the quality of shooting of shortened barrels and when buying

The Teague screw-in choke, with its gradual taper towards the muzzle can be an option for restoring choke in shortened barrels.

old guns, shortened barrels are more likely than almost anything else to dissuade me from the purchase. There is also a balance factor to consider. A quality gun will be made and balanced carefully to produce its fine handling characteristics. Lopping three inches of steel off the end is not going to do anything positive.

Burrard also wrote at length on the role of muzzle flip on the striking point of a gun. The muzzle flip is different according to the length of the barrel, so reducing the length may also affect the point of impact. Typically, a 30" gun reduced to 27" will shoot lower than before it was cut. The practical implications of muzzle flip may be less than the academic analysis suggests but the issue adds another factor to be considered if a shortened gun is offered to you, even more so if you are considering shortening the barrels of a gun that has been fitted to you or with which you shoot well. Such desecration hardly seems credible today, but many such operations were performed in the past and many fine guns ruined because of the vagaries of fashion.

Where there is no choke, the gun may be either shortened or pre-1875 (and therefore possibly best not altered for reasons of preservation and originality). It is always worth remembering that alterations are easier to make than they are to undo. However, many sportsmen continued to have guns made without choke well beyond 1875 or had the excessive choke removed when they realized they were not killing as much game as they used to. Choke can also be lost when barrels are lapped out to remove rust or pitting.

As stated above, pattern your guns and make a record of their shooting characteristics with different loads and if you have several, use those suited to the task when the chance arrives. You can take a lot of satisfaction from bringing out an old gun with cylinder bores for a pheasant or partridge shoot where you know ranges will not be extreme. This is a far better policy than having all your guns multi-choked, in my opinion.

Finally, remember the old adage 'choke will lengthen your range but lighten your bag' and that many great sportsmen shot a great deal of game before choke boring became the norm. Modern cartridges give the modern user of old guns greater options than the original user had – those with enclosed cup wads will produce tighter patterns than the old felt-wadded cartridges. The universal use of crimped turnovers rather than the old paper disc, rolled turnover, method is also a considerable improvement. 'Cartwheel patterns', once the bane of shooters searching

Term	Diameter of barrel (12-bore)	Diameter at muzzle
True Cylinder	0.729"	0.729" (no constriction)
Improved Cylinder	0.729"	0.725"
Quarter Choke	0.729"	0.719"
Half Choke	0.729"	0.710"
Modified (¾) Choke	0.729"	0.701"
Full Choke	0.729"	0.693"

for regularity, have been totally eliminated through the universal adoption of crimped closures.

Choke restrictions are generally explained as a constriction at the muzzle in increments from 5 thousandths of an inch to 40 thousandths of an inch. The terms used and the degree of constriction are shown in the table at the top of this page.

The choke constriction table is a simplified and unreliable measurement of a gun's likely shooting capabilities. Top gunmakers would adjust (regulate in gunmakers' parlance) guns at the shooting ground to achieve performance specifications stipulated when the gun was ordered. Every gun is different and no two guns measuring the same will deliver exactly the same patterns.

Choice of cartridge is also a factor. Guns were often regulated with a specific cartridge and size of shot. For example, I have a Purdey 12-bore and the records at Audley House state that it was regulated for 1oz of No.6 shot when it was re-barrelled in 1973. After experimenting with various loads, the 1oz of No.6 is the one I have settled on for this gun.

The table at the bottom of the page shows typical patterning expectations for a 12-bore gun, shooting 30g of No.6 shot (287 pellets).

As can be seen from the chart, the biggest improvement is from 'True Cylinder' to 'Improved Cylinder'. Each increase thereafter is correspondingly less effective.

Another way of considering choke is by percentage of pellets in a given load, evenly concentrated in a 30" circle at 40 yards. The table above shows the percentages that Purdey regulators work to for each degree of choke:

Degree of Choke	% of pellets in a 30" circle at 40 yards
Cylinder	45%
Improved Cylinder	50%
¼ Choke	55%
½ Choke	60%
¾ or 'Modified' Choke	65%
Choke	70%
Full Choke	75%

For most game shooting 'Improved Cylinder' and 'Half Choke' is a sufficient combination. Many shooting instructors believe most people are probably 'over choked' for general game shooting. Though I have die-hard friends who insist 'full and full' is the best combination for pigeon decoying, I am content with 'Improved and Modified' but maybe that is because it happens to be what I have in my favorite gun!

It is also interesting to note that we have a tendency to find guns choked open in the right (front trigger) barrel and tighter in the left (rear trigger) despite the fact that driven bird shooting usually involves taking the first shot at an approaching bird and the second overhead, thus the second bird is closer and, in fact, in need of less choke.

Perhaps this is historical: when most shooting was walked-up, the first shot was taken close – as the covey rose – and the next bird would be further away when the

Term	Thousandths of an inch constriction	Pellets in a 30" circle at 40 yards
True Cylinder	None	114
Improved Cylinder	3 – 5	145
Quarter Choke	10	160
Half Choke	20	175
Modified (¾) Choke	30	187
Full Choke	40	200

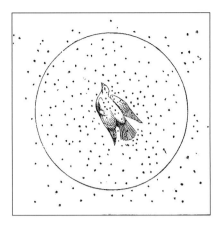

This pattern (left) illustrates the regulation of one of W.W. Greener's 'Pigeon' guns loaded with 1¼oz of No.6 shot.

Right: A modern pattern board used for regulating chokes. This shows an Improved Cylinder pattern being regulated. This shot was taken at 20 yards.

Below: This 12-bore by Thomas Sylven bears 1868 proof marks and I believe is an early live-pigeon gun. Choke boring had not yet been proven through the 'Field Trials' of 1875, so the barrels are not marked 'Choke' or 'Not for Ball'. However, the regulation of the barrels at the pattern plate (done by Army & Navy in 1898) shows patterns averaging at 30 yards (in a 30" circle): 193 pellets left barrel and 190 pellets right barrel. At 40 yards, the figure is 146 left and 150 right. The gun was regulated with Schulze powder and 1oz of No6. As a pigeon gun it would more likely have shot 1¼oz of shot in competition. It weighs 7lb 1oz and is barrel heavy. Even now, the tubes measure a minimum of 41 thou. The 14¼" stock has a straight comb and the rib is wide and concave. It still shoots beautifully.

second shot was taken. Old habits die hard it seems. There is a strong argument for boring game guns 'Improved Cylinder' in both barrels. A friend has a W.R. Leeson boxlock more heavily choked in the right than the left and I have a William Lee boxlock with the same set-up, which was quite commonly applied to grouse guns.

Top extensions and locking bolts

Have you ever wondered what kept your gun closed and locked when it is fired? The whole system works so effortlessly that it is easy to overlook. When coming into contact with guns made in the 19th century, you will find that various ideas were in existence about the best way to lock the barrels to the face of the action and a lot of gunmakers had differing views about the best method and their guns

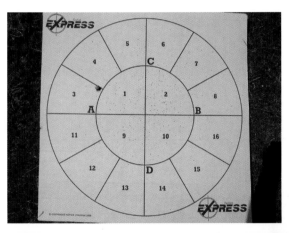

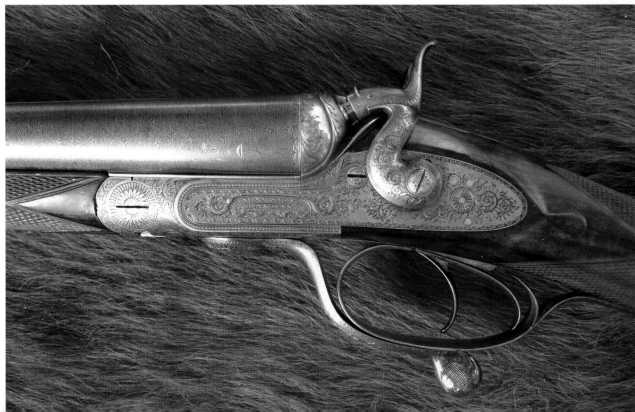

Cocking Indicators

One distinct advantage of the hammer gun is the clear indication that a gun is 'safe' by the visual checking of whether the hammers are at 'cock' or at rest. An un-cocked hammer gun cannot be fired. A hammerless gun may, or may not, be cocked less era. Old photographs of hammergun-carrying shooters often show guns closed at the breech in company, with hammers at rest. Modern etiquette demands an open breech nowadays – an indication of the uncertainty inherent in the safety or otherwise of a closed hammerless gun.

This Pape lever-cocked sidelock has clear indicator limbs showing when the safety catch is applied and tumblers are cocked.

and loaded – the observer cannot tell.

Early hammerless guns commonly featured an attempt to indicate the position of the tumblers. This could be effected by means of a glass 'window', through which the cocked tumbler is visible, or by means of a disc on the side of the action with an arrow or gold-inlaid line on it. When the gun is cocked, the line is inclined, when 'at rest' the line lays horizontal to the bottom of the action. Many sidelocks retain the latter form of cocking indicator but as a practical means of establishing the safety of the gun, it is an ineffective attempt to achieve the clear status of a gun with external hammers.

Gun safety etiquette has evolved into the hammer-

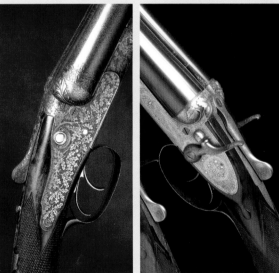

The W&C Scott on the left shows the gold-washed tumbler visible through the crystal window in the lock plate. The Purdey on the right has internal tumblers but the external dummy hammers act as cocking indicators and show the gun as cocked or safe.

featured different systems of getting the job done. If your gun has withstood sustained use until now, whatever system your gunmaker chose, it was a sound one.

There are excellent books available on the plethora of designs for keeping guns closed, and I shall not go into detail here on the more unusual, but the user of vintage guns would do well to recognise some of the better bolting systems likely to be encountered and understand their qualities.

From the time that the breech-loader became a viable proposition, gunmakers and the shooting public obsessed about the potential for such guns to become unsound at the point where the barrels and the action meet. The result was a boom in inventions designed to lock the two together as solidly as possible. In many cases this led to something akin to 'overkill' with makers employing two or three bolting mechanisms, where one would have sufficed.

A famous example of this is the grossly over-engineered Purdey over-and-under with 'sextuple grips', (based on the Edwinson Green design of 1912) which was undoubtedly strong but the appearance and lightness of

The Purdey 'sextuple-grip' over-and-under was very strong but too tall in the action and too heavy. Purdey over-and-under guns are now made on the Woodward system, which Purdey adopted in 1948, when Woodward was absorbed.

the gun was compromised and the multiple grips proved unnecessary. Hardly anybody bought one (only 27 were made) and in comparison with the shallow-actioned, willowy, 1909 Boss and the 1913 Woodward over-and-unders, it is not hard to see why the Purdey failed. However, having handled Purdey guns of this type, I have to say they are extremely good value, wonderful quality and balance surprisingly well.

While the Purdey sextuple grip was somewhat akin to overkill, combinations of locking mechanisms are commonly encountered in breech-loading shotguns. We shall now explore some of those popularly applied patents that the gun trade adopted in the quest for secure barrel locking combined with speed and ease of opening and closing.

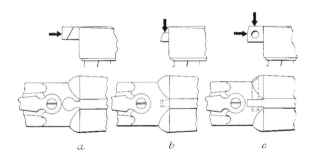

Top extensions (illustrations by Gough Thomas): a) the Westley Richards 'Doll's Head of 1862, b) the Purdey 'Third Bite' of 1868 and c) the Greener patent, showing the 'Wedge Fast' locking cross-bolt of 1873.

Timeline of Developments							
1859	1862	1863	1865	1873	1878	1879	1882
Jones Underlever	Doll's Head Extension	Extension Bolt Thumbhole Lever	Scott Spindle	Greener Treble-Grip	Purdey Third Bite	Rigby/ Bissel Rising Bite	Webley Screw-Grip

The Jones Screw Grip (1859)

The screw grip is an early, strong and simple method of locking the barrels to the action. It operates on a rotary system with a cylindrical head going through the action flat with two grips on an inclined plane. Patented by Henry Jones in 1859 and operated by the rotary under-lever, it quickly became the standard method of closing a breechloader. As the lever is turned closed, the grips lock into the bites on the double lumps under the barrels. Turning the lever the other way pushes the lumps up and disengages the bites to open the barrels. The screw grip is still reliable and was popular in double rifles and large bore wildfowling guns because of its strength. The system looks old-fashioned to modern eyes but should not put off a potential user of a gun on grounds of safety. It is as good as anything subsequently used. Its only drawback was speed of manipulation. Once 'snap-action' designs emerged, they superseded the screw-grip as the preferred operating mechanism for game guns.

Variants on the screw-grip were later patented by Philip Webley and also by James Lang (son of Joseph), both utilizing spring-pressure to make a snap-action. The Lang version was also available as a top-lever operated screw-grip.

The Bolted Doll's Head Extension (1862)

The 'doll's head' – so called because of the shape; a bar extending from the rib into the top of the breech face with a circular top, literally looking like the head of a doll on its neck, was an early system of barrel locking, initially patented by Westley Richards in 1858.

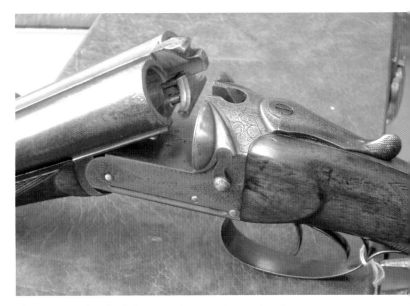

This Newton boxlock features the Westley Richards doll's head and top-lever operated sliding-bolt as the sole method of securing the barrels to the action. Note the individual safety catch for each lock (a feature of the original Anson & Deeley patent).

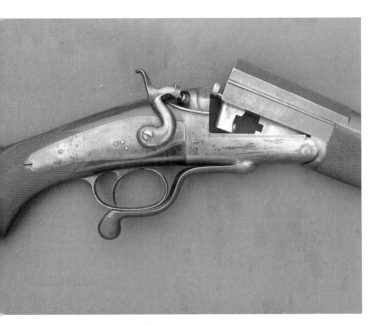

The Jones under-lever and screw-grip of 1859. Greener ungraciously neglected to mention Jones by name in *The Gun and its Development*. However, it remained popular well into the 20th century because of its simplicity and great strength.

This photo of a Westley Richards 12-bore shows a single lump without Purdey bites and the doll's head extension, with a bite which mates with the top lever bolt.

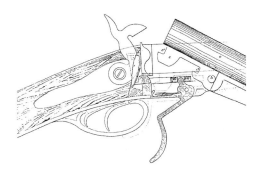

The Purdey Bolt (from a patent drawing). This quickly became the universal means of securing the barrels to the action in breech-loading guns.

The 'doll's head' type of extension of the top rib, operated by a top-lever, was most famously applied by the patentees, Westley Richards, and was initially the sole means of locking the barrels to the action (via a bolt sliding into a recess in the extension itself), following the firm's improved patent of 1862. Once the Purdey double under-bolt became the usual locking method employed, gunmakers often continued to provide a top-extension as added security. Doll's head extensions continued in many second and third grade guns to give peace of mind to the buyer that though inexpensive, the gun would be safe.

The doll's head was also used in quality pigeon guns and wildfowling guns as a means of extra security when firing the heavier loads of shot and powder associated with these sports. There are many variants of the system, many with locking bolts that operate with the top-lever and engage into a bite in the doll's head when the gun closes (as with the Webley screw grip).

The Purdey Bolt (1863, perfected 1867)

The Purdey double under-bolt is the barrel bolting system almost universally adopted by gunmakers nowadays for side-by-side shotguns and double rifles. Systems that preceded it worked well enough and many are strong and practical. However, the emergence of the Purdey bolt in 1863 eclipsed all other methods and has remained the standard to the present day.

The double bolt works by snapping forward into recesses cut into the lumps under the barrels when the gun is closed. This is a very strong method of locking the barrels to the table of the action and does not require any top bolting system to supplement it, though one is often found. The Purdey bolt will be found operated by a number of different levers and all work effectively.

'Treble Grip' Systems (1873)

These systems involve a large, bar extension protruding from between the barrels into the breech face with a hole drilled into the extension to accommodate a locking bolt. The bolt is withdrawn into the fences when the gun is opened and fits back into place when the gun is closed.

Greener's cross-bolt and top extension. In this 'Treble Wedge Fast' action it is used in conjunction with the Purdey bolt.

When done well, it is very neat and an impressive feat of precision engineering.

W.W. Greener invented the system in 1873, initially for use with the Jones under-lever. He called it the 'Treble Wedge Fast' when combined with the Purdey bolt. It is certainly very secure. However, as with much else in gunmaking, the system only works well if the workmanship is good. Some very nasty guns have a version of the 'treble wedge fast' grip and will often work loose quickly. I have only ever found Greener guns to be well engineered and in these cases the 'treble wedge fast' is a lovely thing to operate, though the actions tend to be on the heavy side, at around 7lbs for a 12-bore (Greener felt this the ideal weight but it seems a little heavy to modern tastes, especially in light of the late Victorian and early Edwardian fad for very light game guns).

A variant of the Greener cross-bolt theme is a similar system with a square sectioned bolt (actually a Scott patent known as the 'Improved Bolt' and introduced in 1892') rather than a rounded one. This is not technically as strong, or as neat, though in practice works just as effectively when well made. Many W&C Scott sidelocks and hammer guns were fitted with the Scott 'Improved Bolt'. It was discontinued in 1936.

Another Scott patent treble grip was the 'Triplex Lever Grip' of 1875. This is top-lever operated and features a deep 'doll's head' with a horizontal slot cut into it to receive the top-lever integral bolt. It is commonly encountered on Scott-made guns into the early 1890s (many of these carry the names of other makers as Scott's were important makers to the Trade) and can feature on guns made up until 1914.

The Giant Grip (1877)

The Giant Grip is associated with inventor Henry Tolley and found some popularity as a bolting system for large bore wild-fowling guns. It consists of a large top extension with a rectangular gap into which a top-lever-mounted brace slides when the lever is operated to close the gun. It enjoyed only limited success and was not widely adopted by other makers.

Below: The Giant grip on a 10-bore hammer gun made by Henry Tolley.

The Purdey third-bite of 1878: the extension can be seen above the extractors.

The Purdey 'Third Bite' (1878)

The Purdey 'third bite' (or *Patent Third Fastening* in Purdey parlance) is a neat alternative to the top extension and is less obtrusive, being almost un-noticed in operation. This is a small extension, mid-way between the barrels, which is locked into place by a bolt acting on the Scott spindle. It is used in addition to the conventional Purdey double bolt. It has the advantage of not protruding from the top of the action and does not interfere with quick reloading; a charge that many have made against the doll's head and other large extensions, such as those common on guns with Greener-type bolting mechanisms. Purdey still offer it as an option.

Rigby & Bissell's 'Rising Bite' (1879)

This is a complex and beautifully engineered means of securing the top extension to the breech face. In this system, a top-extension with an elongated hole machined

A snap under-lever sidelock by Rigby featuring the Rigby & Bissel rising-bite.

into it is engaged by a rising-bolt, operated by the snap action under-lever. Pressing the lever causes the bolt to retract downwards. Closing the gun is effected in the usual manner of raising the barrels and the gun snaps shut. Used in conjunction with the Purdey bolt.

The Webley Screw Grip (1882)

This was patented in 1882 by Webley and Brain and is similar to the Westley Richards 'dolls head' extension in appearance, but with a flat extension 'table' below the 'doll's head' which is engaged by a threaded spindle, worked by the top-lever. The top-lever acts on the Scott spindle and Purdey bolt in the now conventional manner. When the top-lever is turned to the right, the Purdey bolt is withdrawn and the threaded spindle disengages with the top extension.

The Webley Screw Grip on a boxlock ejector by J. Woodward c.1910.

When the gun is closed, the Purdey bolt engages the lumps and the threaded spindle secures the 'dolls head' extension. Thus, the action is secured from the bottom and the top. This type of grip can be found on Webley guns dating from 1882 until 1946; hammer guns, boxlocks and sidelocks (generally built on the Rogers seven-pin sidelock action) with various names on the lock-plates may be encountered with the Webley screw grip, as Webley made guns for many other firms. Notable among those retailing Webley guns with the screw grip were Army & Navy, William Evans and Trulock & Harriss.

When buying vintage guns, it is not wise to worry too much about the locking system employed. They are historically interesting and ingenious devices that form part of the history of gunmaking but a Purdey bolt on its own is sufficient for almost all circumstances, as long as the gun is well-engineered and of quality materials. Earlier guns using the Jones screw grip are perfectly sound and serviceable and even those using the 'doll's head' alone have given good service for long enough to prove their reliability.

Operating levers

To make the bolting system open and close, a lever of some kind must be attached to it. Manual manipulation of this lever will cause the bolt to move from the locked to the unlocked position and allow or cause the gun to open.

The Top-lever (1858)

The top-lever, introduced by Westley Richards in 1858 and improved in 1862, 1864 and 1871 is the most common type of opening lever currently in use. In the original patents the

The top-lever used in conjunction with the Scott spindle and the Purdey bolt, shown here on a Westley Richards boxlock, is the most favoured combination of operating lever and bolting device.

top-lever operated a bolt which engaged with a slot in the 'doll's head' extension, though it generally functions most successfully in tandem with the Scott spindle (since 1865) and Purdey double bolt. The lever is operated by using the thumb to push it to the right. This movement turns the spindle, which slides the Purdey bolt back, out of the bites in the lumps, and allows the gun to open. The combination of top-lever, Scott spindle and Purdey bolt has become the standard method of opening and closing a breechloader. Millions have been made.

The Rotary Under-lever (Jones Underlever) (1859)

The rotary under-lever is a lever wrapped over the trigger guard with a knob protruding at the rear. This is pushed to the side to operate the bolting system. It must be pushed back into the closed position manually once the barrels have been closed to the action face. The rotary under-lever is commonly associated with the Jones screw grip, described and illustrated earlier.

This system is inert and slower to operate than the 'snap-action' bolts and levers but it is a reliable and very strong method of securing the barrels to the action-face and, as well as being found in early shotguns, was retained as a method of securing heavy double rifles and large-bore wildfowling guns. I recently saw a massive double 8-bore hammer rifle by H. Holland, incorporating the Jones screw-grip and rotary under-lever. If it was good enough for such a monstrous cannon, then it surely had nothing more to prove. Sadly, Jones did not have the money to make the most

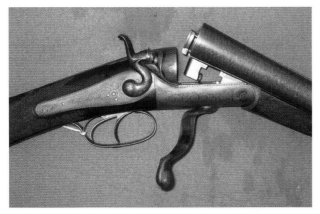

This back-action Holland & Holland 10-bore hammer gun shows the Jones under-lever in the 'open' position.

of his patent and died a poor man despite having invented one of the classic methods of locking a breechloader.

The Horsley Sliding Top Bolt (1863)

The top-lever on some guns following the 1863 Horsley Patent (and others) for the sliding single bolt superficially resemble the more normally encountered top-lever, which

Below: This Horsley gun shows the sliding top-bolt. Rather than pushing the lever to the right, one pulls it back to disengage the bolt from the bite in the lump.

is pushed to the right. However, operation is effected by thumb pressure moving the lever backwards and disengaging the locking bolts.

The 'Lift-Up' Top-lever (1864)

Another variant is the William Powell 'Lift-Up' top-lever of May 1864. The lever in this case is lifted with the thumb to perform the same opening operation. Both these methods are perfectly robust and convenient to operate. However, they did not become popular and once the top-lever and Purdey bolt were combined, the Horsley and Powell patents faded into disuse.

This William Powell top-lever is pressed up to disengage the sliding top bolt.

The Scott Spindle (1865)

In 1865, William Middleditch Scott patented the link necessary to combine the practicality of the Westley Richards top-lever with the security of the Purdey double under-bolt. Scott's spindle runs horizontally through the action, from the top-lever to the Purdey bolt, which it operates via a cam. This classic combination of these three patents has been the most widely adopted means of opening and locking a side-by side shotgun ever since.

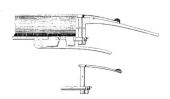

The Scott spindle was operated by a top-lever. This has become the 'default' method of opening a breech-loading shotgun.

The Side-lever

The side-lever works in the same way as the top-lever but is simpler, being an extension from the Purdey bolt to the chosen side of the lock plate. The lever is pressed down to open the gun. Side-levers can be equally easily made to operate right or left-handed. They appear on guns by many makers but seem to have been a particular hallmark of Stephen Grant, both for hammer and hammerless guns. Boss used the side-lever more than any other variety in the hammer gun era but abandoned it for the top-lever & Scott spindle in the 1890s, as did most other firms.

A Grant bar-action hammer gun with side-lever. The side lever is particularly common on best-quality guns by Stephen Grant and hammer guns by Boss.

The Thumbhole Lever (1863)

This works on the same principle as the snap action under-lever but the lever is actually incorporated into the trigger guard. It acts directly on the Purdey bolt to open the gun. It is the original method employed with the Purdey bolt and dates from Purdey's patents of 1863. Further improvements were made to this but they all look very similar and work on the same principle. Thumbhole levers vary quite widely in the shape of the lever and fairly free experimentation with the shape and form of the lever itself appears to have taken place at individual gunmakers'.

Purdey's thumb-lever is shown here, on a back-action hammer gun by Reilly.

The Snap Action Under-lever

This is another variant on the original thumbhole lever and usually takes the form of a lever extending over the trigger guard. It is pushed forward to draw back the Purdey bolt.

A Rigby 12-bore with snap under-lever mechanism.

Pape's thumb tab on a gun bearing his name.

Closing the gun causes the lever to spring back into place, with no further manipulation required to lock the gun shut. Since the bolt is under tension from a spring, it locks immediately into the bites in the lump. Hammer guns with snap under-levers exist (like the 1878 Hughes self-cocking hammer gun) but they became more widely used during the 1870s, when hammerless guns began to emerge. Most of these early guns were lever-cocked and the snap under-lever was a popular device for combining the opening and cocking operations.

The Push-Forward Thumb Tab

WR Pape of Newcastle patented this method of opening and closing a gun in 1867. The tab under the bar of the action pivots and engages with a single bite in the under-lump. It is a snap-action and appears quite sound despite the rather flimsy-looking bolting system. I have only ever seen it used on hammer guns made by Pape.

Value and Popularity Issues

Prices of guns are certainly affected by the operating mechanism. Most buyers want guns with a top-lever, reflecting the current norm. This is true for both hammer and hammerless guns. Side-levers are next in line in popularity and the snap action under-levers follow. Rotary under-levers, slower to operate and more 'old fashioned', are the

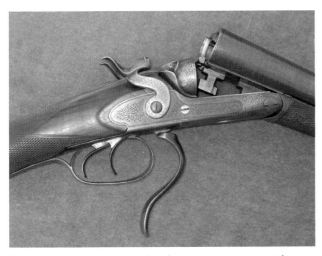

The unusual can prove to be the most interesting. This Lang (E. Hughes 1878 patent) features a snap-action variant of the rotary bolt and automatic lever-cocked hammers.

least popular. I, however, like them and as long as driven game in high volume is not the order of the day, I don't feel overly handicapped by using one. Thumbhole levers are also a feature that can lower the sale price of a gun of fabulous quality. They are getting more popular as this becomes apparent and prices are creeping up.

The forend

The forend (or fore part) of a shotgun serves to hold the barrels to the action and also houses the ejector mechanism in many guns. Traditional British side-by-side gun design never considered the forend as having a primary role in the holding or pointing of the gun. It is usually of a 'splinter' style. Most shooters, for many shots, will hold the gun in front of the forend, actually gripping the barrels.

American and modern continental side-by-sides often feature a fuller forend; this is designed to give grip and protection to the hands from hot barrels and is usually termed a 'beaver-tail forend'. Some vintage guns will feature these as later additions or special orders.

Some vintage guns of ultra-lightweight design will have tiny forends. This feature is mostly found in lightweight hammer guns (Turner made a well-known version called the 'Featherweight').

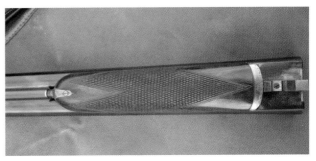

The Anson pushrod of 1872.

Methods of attaching the forend

The most commonly adopted method of removing the forend on a side-by-side nowadays is the 'Anson push-rod' (1872), type. It is easy to remove, reliable and neat in appearance. The forend is removed by pressing the tip of the rod and releasing the spring-loaded bolt that engages a bite in the barrel loop. It is replaced simply by pressing into place and works as a 'snap action'.

Deeley & Edge forend fastener of 1873.

The Deeley and Edge (1873) patent features a lever sunk into the wood and is operated by hooking a finger into the recess and pulling downwards. I find this the most elegant and attractive, as well as the easiest to manipulate

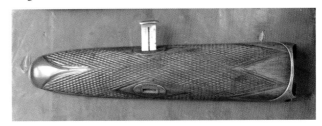

This wedge and escutcheon is common on muzzle loaders and early breech-loading guns.

with cold fingers but it requires the fitting of more wood-to-metal parts and is probably more time-consuming to produce. It is also a 'snap action' type.

Older guns may feature a metal wedge that is pushed through a slot in the side. This is strong and

Snap-action forend fastener; there are many variants. The 1878 Hackett is perhaps the best known.

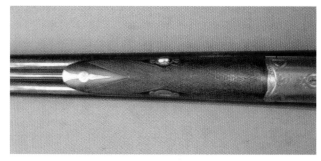

A push-button side-catch on a Woodward hammer gun.

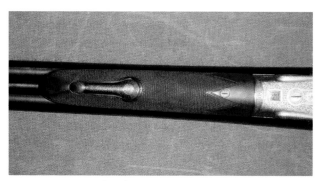

Harvey 'grip catch' of 1866.

Yet another variant; this one from a Pape hammer gun.

The Scott forend catch of 1876.

foolproof (dating back to muzzle loading days) but needs a metal-edged tool to remove, unless you don't mind broken thumbnails.

Other forends are simply press-on/pull-off and are retained by a spring clip. The 'Hackett Snap' (1878) being the commonly encountered version. The 1866 Harvey 'Grip Catch' is an inert but reliable fixing system that will occasionally be found on double shotguns but is more often encountered on double rifles.

The British gun trade was full of invention and one is constantly surprised by the discoveries of old inventors. I recently encountered a forend secured and released by pressing a button on the side to disengage the locking bolt. I had never seen this before and have not seen one since. Doubtless there are other such curiosities waiting to be found.

When buying a gun with sleeved barrels, check the fit of the forend and the regulation of the ejectors carefully. The new tubes are difficult and time-consuming to 'strike up' to the dimensions of the original ones and it is not uncommon to find guns with problems in this area. These are easy to overlook but can be very difficult to remedy without a lot of time and expense. Be warned; it can be as expensive to re-fit new wood to a forend as to replace the buttstock of some guns. I recently asked a stocker to quote for such a job on a best gun and the quote was $2,000!

Proof

If you are going to buy old guns and use them, you need to understand whether they are safe and in suitable condition for use. This can be daunting when you get started; appraising a vintage gun without knowing what to look for is like buying a second-hand car without having any mechanical knowledge. Just because it is shiny and has a nice badge on it does not indicate that it will be reliable or safe to use.

Luckily for us, the British have laws, dating back to 1813, relating to the sale of guns that go some way to protecting the public from unsafe guns of any age.

All British guns are tested before they can be sold by a system known as 'proofing'. This takes place in one of the UK proof houses: Birmingham or London. To put it simply, new guns are tested by having much greater loads fired through them than they will be expected to actually fire in normal use. If they withstand this pressure, they can be assumed safe with the lower pressures of the nominated cartridge size for which they are considered 'proved'.

When considering a breech-loading shotgun of age, it will have been proved at least once and very likely more than once (in which case the re-proofing marks will be evident alongside the original proof marks).

1868 London black powder proof marks on this hammer gun help confirm the completion date of the gun, which has many features of guns made four or five years earlier.

states, 'proof' is just that, it is proof that the gun will safely shoot a given load. Anything else is a theory or a guess and the system of seeking proof that an arm is safe is the better one. It is still possible to have a gun re-proved for black powder, though there seems little point.

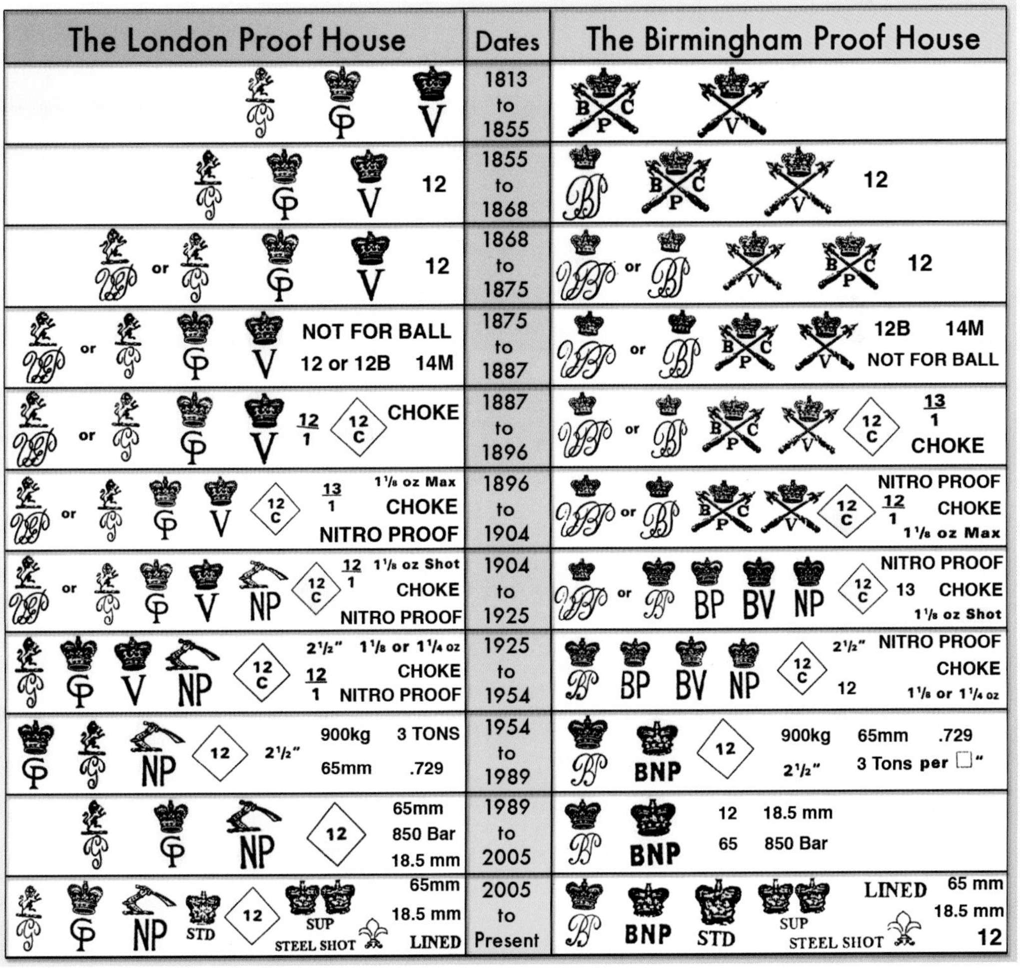

Black Powder Proof

This means that the gun has been proved for use with cartridges loaded with black powder only. It is not safe to use it with modern nitro cartridges. For such guns it is possible to buy specially loaded black powder cartridges and use these exclusively, or submit the gun for re-proof.

Some American writers have experimented with the pressures produced by black powder and nitro powders and concluded that it is safe to fire nitro loads with pressures equal to those produced by the equivalent black powder load.

While there may be merit in the exploration of the truth of the black-powder/nitro load performance it would be irresponsible to advise a user to put nitro shells in any gun not nitro-proved. As the term

The London Proof House	Dates	The Birmingham Proof House
(proof marks)	1813 to 1855	(proof marks)
(proof marks) 12	1855 to 1868	(proof marks) 12
(proof marks) or (proof marks) GP V 12	1868 to 1875	(proof marks) or (proof marks) 12
(proof marks) or (proof marks) GP V NOT FOR BALL 12 or 12B 14M	1875 to 1887	(proof marks) or (proof marks) 12B 14M NOT FOR BALL
(proof marks) or (proof marks) GP V 12/1 12C CHOKE	1887 to 1896	(proof marks) or (proof marks) 12C 13/1 CHOKE
(proof marks) or (proof marks) GP V 12C 13/1 1⅛ oz Max CHOKE NITRO PROOF	1896 to 1904	(proof marks) or (proof marks) 12C NITRO PROOF 12/1 CHOKE 1⅛ oz Max
(proof marks) or (proof marks) GP V NP 12C 12/1 1⅛ oz Shot CHOKE NITRO PROOF	1904 to 1925	(proof marks) or (proof marks) BP BV NP 12C NITRO PROOF 13 CHOKE 1⅛ oz Shot
(proof marks) GP V NP 12C 2½" 1⅛ or 1¼ oz 12/1 CHOKE NITRO PROOF	1925 to 1954	(proof marks) BP BV NP 12C 2½" NITRO PROOF 12 CHOKE 1⅛ or 1¼ oz
(proof marks) GP NP 12 900kg 3 TONS 2½" 65mm .729	1954 to 1989	(proof marks) BNP 12 900kg 65mm 2½" .729 3 Tons per ☐"
(proof marks) GP NP 12 65mm 850 Bar 18.5mm	1989 to 2005	(proof marks) BNP 12 18.5 mm 65 850 Bar
(proof marks) GP NP STD 12 65mm 18.5mm SUP STEEL SHOT LINED	2005 to Present	(proof marks) BNP STD SUP STEEL SHOT LINED 65 mm 18.5 mm 12

Nitro Proof

Once black powder was replaced in general use by nitro propellants (Schultze produced his powder in 1866 and 'E.C' followed in 1882 with Nobel's 'Ballistite' in 1888 cementing the nitro revolution), new proof tests were required to ensure that guns were able to withstand the pressures of modern loads (from 1887 the proof marks will show the powder and powder load – e.g. SCH 42 1⅛ for Schultze powder, 42 grains, 1⅛oz of shot). Many guns originally proved for black powder will have subsequent proof marks from when they were submitted for the new 'nitro-proof'. They will be stamped with the 're-proof' mark as well.

Dating Guns by Proof Mark

Because proof marks have been used for hundreds of years in Britain and because they have been updated and changed over the years, it is possible to help discover the age of a gun by deciphering the proof marks stamped on the barrel flats.

The tables (see page 139) indicate key dates for changes in the marks stamped by the proof houses. By referring to the time line and the proof marks on any vintage gun, it is possible to discover the date of the earliest proof mark stamped on the gun and therefore narrow down the age of the piece.

Marks vary between the two proof houses so it is easy to discover whether a gun was proved in London or Birmingham. This is not as reliable as it may seem: when seeking information about a gun which has a maker's name but no address, the London proof marks do not necessarily indicate that you have a 'London' gun. Many provincial gunmakers sent barrels to London for proof because of the cachet of the London association. Many 'Birmingham' guns will likewise have a London address on the rib or lock plates because the maker had a London retail outlet. Also, many 'London' guns were actually made in Birmingham before being sent to London for proofing and 'finishing' so your London-proved barrels may easily have been made in Birmingham: in fact, most were.

Confusion

It is not uncommon, as previously stated, to encounter a number of proof marks on a gun. It is important to be methodical and work through them systematically in order to date your gun. The earliest proof marks will indicate its original submission to the proof house but they may well be mingled with later marks. It is usually possible to arrive at a date before which the gun could not have been proved. Dating guns is an inexact science and is often a combination of eliminating possibilities until you reach a period of a few years that you are sure mark the earliest and latest possible for the manufacture of your gun. This detective work is part of the fun. Consider also the distinctive features of the gun's design, for many guns will be encountered that did not see the inside of a proof house until a long time after they were started.

Provisional and Definitive Proof

Barrels were submitted for provisional proof before work had gone 'too far' on the tubes. This process tests them for basic soundness before more expensive time and energy is spent on them. Any flaws in the metal are exposed and unsound tubes rejected. This is less of an issue with modern production methods but in the days of Damascus barrels it could save gunmakers a lot of money in wasted time finishing barrels that would likely fail later.

Definitive proof is carried out on the barrels and the action by firing an overly heavy charge to check the soundness of both. If proof is passed, the gun is finished and released as safe for use.

That was then, this is now. Your gun may have been passed as 'safe' by the proof house when it was made 100 years ago but much time has passed and a lot may have happened since. The proof laws protect us in such cases because the dimensions are set at original proof. If the gun exceeds the original dimensions by ten thousands of an inch (eight thousandths of an inch once metric marks are used) it is deemed 'out of proof' and may not be legally sold in Britain. There are other reasons that a gun may be deemed 'out of proof', all relating to weaknesses in the action and barrels. If a gun is materially altered by procedures such as the addition of multi chokes, it will need to be submitted for re-proof.

This information is not much use unless you have a way of reading the original dimensions and measuring the current ones. When buying at auction, the auctioneer must provide measurements of bore size and barrel thickness. When buying from a dealer, the measurements will also be available and should be checked in your presence. When buying from a private individual, take it to someone competent to measure the barrels before agreeing on a price unless you are able and willing to go on intuition and take a risk because 'the price is right'. If in doubt, check.

It is reasonable for the layman to wonder what all the fuss is about. What can go wrong with a steel tube? Well, neglected barrels will rust and internally, this rust can become 'pits'; pits are areas where the rust eats a hole into the bore. These need to be removed if the rusting is to be stopped (a process known as 'lapping'). Lapping removes metal from the inside of the tubes and widens the internal dimensions of the bore. This weakens the barrels and, if done beyond ten thousandths of an inch, will render the gun 'out of proof'. This means that it will need to be submitted for proof again – and it may not pass (i.e. it may blow-up and leave you with a heap of twisted metal instead of a gun).

In practice, most guns unlikely to pass proof will be stopped before they reach the proofing stage by the practiced eye of the viewing officer. However, I was recently in the gun-room of Atkin, Grant & Lang and saw a nice sidelock action that had cracked through the right-angle between action flat and table in the proof house, so the venture is not without risk.

It is also worth remembering that when British guns were made, they were usually made specifically to the demands of their original design: hence the light, fine actions on lightweight guns with 2½" chambers designed for light loads. By extending these to 2¾" and putting more powerful shells in, the action will be subject to more stress than it was intended and may not stand up well under this abuse.

Useful figures

• **Ten thousandths of an inch.** If your barrels exceed this internal size in excess of the dimensions measured at proof, it is deemed 'out of proof' and is illegal to sell in the UK.

• **Twenty thousandths of an inch**. Be very careful if you

are thinking of buying a gun with walls thinner than this. The trade prefers to see twenty five thousandths of an inch in the walls of the barrels. Very thin barrels may be prone to dents and raising these dents could make the barrels unsound.

Chambers and Proof

Most modern shooters know that cartridges come in different sizes. The usual sizes are:

2½" (65mm & 67.5mm) Most common in vintage guns.

2¾" (70mm) Usually only in old wildfowling or pigeon guns.

3" (75mm) Sometimes encountered in wildfowling or pigeon guns.

Guns with 2½" chambers will physically take 2¾" (70mm) cartridges but **must not** be used with them. The important consideration is not the length of the cartridge but the pressure exerted when it is fired. Many nominal 70mm cartridges are loaded with light shot loads but are not to be considered safe in 2½" stamped guns. Even a 24g load in a 70mm marked cartridge will produce heavier pressures than a 2½" stamped tube is proved safe for. Do not be misled by the idea that an old gun stamped with the proof mark 1⅛ is safe for use with modern 70mm cartridges with this load of shot. The proof house used standard powder loads with each shot load – modern 70mm cartridges produce higher pressures.

Typical proof pressures	
2½" Cartridge (65mm 67mm or 67.5mm cases)	3 tons per square inch
2¾" Cartridge (70mm cases)	3¼ tons per square inch
3" Cartridge (75mm cases)	4 tons per square inch

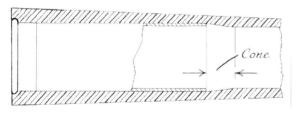

This diagram shows the breech end of a shotgun barrel, with cartridge inserted. The chamber is sized to fit the cartridge and the cone leads the ejecta into the bore. Guns with tight cones and 2½" chambers may tear the crimped ends of modern 67mm cartridges. If this occurs, try true 65mm cartridges or have the cones eased.

Make sure the information on the cartridge box indicates suitability for your vintage gun. Note the Eley Maximum is a 34g shell suitable for use in 2½" chambers. The Gamebore Pure Gold is also made for 2½" chambered guns but carries only 28g of shot. Also note the different case lengths suitable for nominal 2½" chambers: 67.5mm for the Eley and 65mm for the Gamebore

2½" chambered guns with tight forcing cones may shred the ends of cartridges labelled as safe for their chambers but 67mm or 67.5mm in actual case length. This adds to the strain on the action and may increase recoil. If this occurs either have the forcing cones eased or stick to true 65mm cases.

It is worth the effort to study the various proof marks and their corresponding dates. It will help you trace the history of your gun as well as assess its fitness for use. Useful information and informative tours are available from the proof houses:

> The Proof Master
> The Proof House
> 48 Commercial Road
> London E1 1LP

> The Proof Master
> The Gun Barrel Proof House
> Banbury Street, Birmingham B5 5 RH

Reproofing

When a gun is encountered that is not currently useable because it is 'out of proof' it may be re-submitted for proof at either proof house. The proof house charges in 2007 are around $87, including VAT. Do not dismiss this as an option. The process is straightforward and can be arranged by any competent gunsmith. Talk to him before deciding if the gun should be submitted for proofing. His advice will usually be sound – but the risk is yours. There is always a danger that an old gun will fail in the action, as well as the barrels, when proof loads are put through it.

When a gun is submitted to the proof house for re-proof it will first be subjected to a visual examination by a proof house officer. If the gun is materially damaged, unsound or in poor condition it may be rejected as unsuitable for proof. It is advisable to have the gun prepared for proof by a competent gunsmith before it makes the journey to the proof house. He will make sure it is 'on the face' and remove any pitting before releasing it.

Below: The Churchill Premiere XXV is a wonderful gun for shooting fast partridges or flushed pheasants. Don't discount the 25" gun just because of the current fashion for long barrels.

Barrel length

I detect a definite trend towards longer barrels among shooting companions. Thirty inch 20-bore and 28-bore over-and-under guns are now generally offered by Beretta *et al* and the auction scene has shown that the once 'old fashioned' 30" tubes are back in fashion with buyers of vintage guns. It is interesting to explore the history and the myths surrounding barrel length and the debate that has raged over the years.

The most commonly-requested barrel length for a side-by-side shotgun has for some time been 28". However, a debate regarding the merits of various barrel lengths obsessed shooting men in the first half of the 20th century and it is still an issue that is hotly contested today. Everybody seems to have an opinion, some based on sound theory, others on myth and misapprehension. Here are some of the opinions of well-known writers on the subject over the years.

I'll begin with the prevailing view when I was growing-up as expressed by MacDonald Hastings (who was also Robert Churchill's ghost writer) in *The Shotgun* in 1981:

'No doubt the length of gun barrels advocated by our forefathers was excessive. At 32" and over there is a downward flip in the metal of the tubes, which encourages the common fault of shooting behind a moving target. At 25" there is an upward flip in the metal.

The gun, because of the shorter length, lifts quicker onto the target, especially in covert, but makes a recognizable crack on discharge and, undoubtedly, in hands inexperienced in taking recoil it has a sharper kick. When 25" barrels were first introduced, it was also said there was a loss in the velocity of the charge.

So there is, but it is negligible. Almost all game is shot well within the limits of any gun, whatever its barrel length. The vast majority of shotguns built in London today have barrels of 28" which experience has shown have neither a downward nor upward flip. Aesthetically a tall man looks better and possibly shoots better, with barrels of that length'.

Robert Churchill championed the 'XXV' short-barrelled game gun. His advocacy of a short-barrelled, lively gun and an instinctive snap-shooting technique was very controversial. After decades of argument, the current fashion has swung again towards longer barrels.

143

Hastings himself was a tall man but shot with 25" barrels. He rightly observes that the norm in early fire-arms was for long barrels and this held sway for as long as black powder was the main propellant in use for shotgun cartridges, though it is not true that black powder needs a full 30" to deliver its full effect, as has often been suggested. With the advent of faster-burning smokeless powders shorter barrels began to enter the consciousness of the shooting public. However, the short-barrel movement was largely championed by one man; Robert Churchill.

Legend has it that Churchill, a gifted competitive shot, was on his way to a shooting match when his gun was damaged in transit. Too late to have new barrels fitted: he instructed his gunsmiths to cut the barrels down and recess-choke them. He shot so well with the truncated gun that he was converted to the concept of short, light 12-bores and the 'Churchill XXV' was born. This is a nice story but Churchill himself vigorously denied it and actually put a great deal of thought into developing the whole XXV package, not solely the length of the barrels.

The magic 'XXV' marketed by Churchill from just before WW1 (with the phrase *why wait to see what a light short-barrelled gun can do for your shooting?*) succeeded in fixing this length in the mind of the public but barrel lengths of 26" and 27" were also made in good quantities. In 1930 Holland & Holland devised their 'Royal Brevis' – a 'Royal' sidelock with 26½" barrels – to meet the trend. Cogswell & Harrison made something of a speciality of 27" barrels.

Some critics believed, erroneously, that shorter barrels did not shoot as 'hard' as longer ones. (i.e. they did not deliver the shot charge with the same velocity). This is not the case, as we will hear from two eminent authorities on the shotgun, Tom and Jim Purdey, writing in 1936:

The shooter may ask if a 27" length barrel shoots as good a pattern as a 30" or a 29" and the answer is that the difference is so infinitesimal as to be not worth worrying about, he will get almost identical results. From the shooter's point of view it might be well to bear in mind that a 25" barrel should be choked a little more than the 29" or 30" barrel… the penetration, that is the killing power of the guns, will be very much the same out of both guns'.

Fifteen years later the point was reiterated by Major Burrard: *'There is no general best length of barrels. The best length of barrels for any individual is the length with which he finds he can shoot best. Barrels of 25" will prove just as killing as those of 30"…'*

So much for the gunmakers and ballisticians,

what of the shooting instructor and great clay and game shot Percy Stanbury? Here is his view, penned in 1962: *'The average man of about five foot ten inches is generally suited by a 28" 12-bore and the six foot fellow with a long reach can well take 30" barrels. Long barrels reduce muzzle blast and recoil and help towards steady shooting.'*

GT Garwood took a pragmatic line in 1969, when he reviewed the Churchill influence: *'All he (Robert Churchill) really established was that there are good practical reasons for barrels shorter than 28", subject always to personal preferences….if only a man is confident in that he shoots better with barrels that are shorter or longer than the common choice, or even if he merely prefers them, he is beyond the need or reach of argument.'*

Other writers felt there was a correlation between bore size and barrel length. H.A.A. Thorn, W.W. Greener and Burrard all had something to say on the subject. Thorn, in the final edition of Charles Lancaster's *The Art of Shooting* (1937) suggested a rough guide: *'Since 1920 the majority (of barrels) have been 28" long, although more than this length is often required and many hold shorter barrels in favour, as quite a number did in 1875'.* This applies to 12-bores.

Thorn echoes others in the assertion that bore size was a significant determiner, generally towards the need for shorter barrels: *'In smaller bores, 28" barrels are often felt to give a 'billiards-cue' effect and 26" is the preferred length.'*

Greener, well known for his '96-1' weight guide for shotguns and their shot charges seemed to have a definitive formula for almost everything. Barrel length was no exception. In *Modern Shotguns* (1887) he tells us: *'The proportionate length will be ascertained from the ratio of length to calibre – 40-1 holds good for shotguns…with chambers of the usual length (2½") the 12 gauge choke-bore barrel is better under than over the 29.16" which is theoretically its correct length. Barrels of 28" seldom fail to give complete satisfaction but the short barrels should not be chambered for extra-long cartridges* (by which he means 2¾").

Burrard also liked a formula and suggests the following as a guide:

'…very large bores will need a greater minimum length of barrels than very small. As a general rule it can be taken that 10-, 8- and 4-bores need average lengths of barrels which are 2, 4 and 6 inches longer respectively than the corresponding average length for an ordinary 12-bore. On the other hand, smaller sizes than twelve can be fitted with barrels that are, on average, slightly shorter. For example,

A 1930s Churchill 'Premiere' grade 12-bore XXV built on the 1906 Baker back-action.

Churchill's shooting style and his short-barrelled XXV guns work well within 30 yards. Many prefer longer barrels for a steadier swing on really high birds.

neither a 16-nor a 20-bore should ever be fitted with barrels longer than 28 inches.'

Writing in 1947, Burrard's opinion is at odds with the 2004 practice of equipping small bore guns with 30" or even 32" barrels. Long barrelled small bores were also relatively commonplace in the 19th century but this is partly explained by the use of black powder at the time. My own Holland & Holland 16-bore feels perfectly well suited to its 30" barrels, as originally fitted in 1885.

The debate goes on. Buyers of vintage guns bring their own ideas about barrel length to the saleroom and over the years fashion and trend cycles have waxed and waned.

Shortened barrels on 19th century guns are reminders of the Churchill-inspired rush towards the XXV ideal and the perceived obsolescence of the traditional 30" tubes. Major Burrard deals with several requests for advice about shortening barrels in his book *In the Gun Room* and Richard Arnold in *The Shooter's Handbook* warns readers against trying to shorten barrels at home; it was clearly a hot topic in the early 1950s, when these books were penned. Guns with shortened barrels should be viewed with care – most or all of the choke will have been lost and the balance will not be as originally intended by the gunmaker. Shortening barrels has a detrimental effect on prices realized.

Generally 'live pigeon' guns and wildfowling guns will have longer barrels than game guns and pre-1890 12-bore guns would normally have 30" barrels, though Greener wrote of some customers requesting barrels as short as 24" as early as 1875, echoing Lancaster's observations. A friend has in his collection a Rigby double 12-bore muzzle-loader with 24" barrels, recorded as original in the Rigby register for 1847. At the other extreme, Greener and Webley exported 'colonial grade' boxlocks, for retail by local stores like Rawbone in South Africa, with 36" barrels. These were reputedly for shooting from horseback. The long barrels helped stop the rider from shooting his horse in the back of the head!

The resurgence in interest in longer barrels means that 30" guns are becoming more sought-after at auction, perhaps even more than 28" guns. This is just fashion.

I learned to shoot with a 12-bore with 30" barrels and I have a 16-bore hammer gun with 30" Damascus tubes that handles beautifully. However, my usual game gun is 28" and I have another with 25" barrels. If I ignore the length of the barrel and concentrate on the bird, I really find there is little difference in performance. I prefer slightly shorter barrels in a pigeon hide and the longer ones for driven game but this is largely down to the space available to swing the gun.

Robert Churchill coached a style of shooting that involves a fast mount and shoot rhythm with an 'instinctive lead' picture and the shifting of weight to left or right foot according to the shot to be taken. This style is what I generally use in the pigeon hide as it minimizes movement and provides the most flexibility in a confined space with targets at unpredictable angles, visible for a short time.

When I tested a 1939 Churchill Premiere XXV 12-bore, built as a 'best' gun on the 1906 Baker sidelock action, I was immensely impressed with the ease with which fast mounting and shooting was possible on fast, driven partridges. I shot very well with the gun and was struck by the whole concept of the Churchill XXV and the way that the entire gun was built around Robert Churchill's principles. It was a wonderful example of how the very best guns are so much more than the sum of their parts.

The Stanbury method I favour for game shooting works (for me) better with longer barrels. There is more space to set the feet as they are best placed to take each shot and a fuller swing ensues. However, professional pigeon guide Will Beesley favours very long barrels in a pigeon hide, so it appears to me to be more a matter of personal taste than anything more certain or stylistic.

Long v. Short Barrels: to sum up...

My own conclusion inclines towards the theory that the psychological effects of barrel length outweigh practical issues.

Longer barrels may make for a steadier swing and shorter barrels may get onto the target faster. Neither choice offers any advantage in delivery of the payload – a longer barrel will not increase your range.

Like choice of cartridge, (and, to a degree, choke) the mindset of the shooter and his confidence in his equipment is of primary importance.

Live pigeon guns

The sport of live pigeon shooting was popular in England from the early 1800s until 1921, when it was finally banned under the Captive Birds Shooting Act of that year. It continued abroad, notably on the French Riviera and at Monte Carlo. Long after having been banned in the UK, it still takes place at a few select locations on the continent and in the USA and parts of South America to this day.

The sport demanded a different type of gun to the usual game guns produced for general sporting purposes and a wide range of makers competed to meet this demand. Pigeon shooting attracted a great deal of betting and many of the competitors were very wealthy individuals. The sport attracted a deal of press interest, scores being reported much like the results of football matches or motoring Grand Prix.

Basically, pigeon shooting required the shooter to drop his bird (generally a blue-rock pigeon) within the confines of a perimeter fence, otherwise it was considered 'lost'. The pigeons were released from a trap from 16 yards to 35 yards from the shooter; rules varied from club to club as the sport developed. The shooter wanted to plaster his bird with as much lead as possible to ensure a 'kill', since the loss of one bird could cost him the match. As a result, a specialised type of gun was developed to meet the demands of the sport.

The gun collector will encounter many guns that can be loosely described as 'live pigeon guns' from percussion muzzle-loaders to modern sidelock ejectors. Pigeon guns generally have one or more of the following telltale signs that differentiate them from their game shooting cousins.

Weight

Pigeon guns tend to be heavy. The renowned pigeon shot Captain Adam Bogardus preferred a W&C Scott hammer gun weighing over 10lbs, whereas his rival 'Doc' Carver used a Greener weighing a little over 8lbs. When gun club rules were generally established, guns were standardised at a maximum 7lbs 8oz at The Hurlingham Club and 8lb at The Gun Club, Notting Hill. When encountering vintage guns, this kind of weight is a clear indication that the gun was originally intended for live pigeon shooting. Greener, in 1891 observed *'The gun for pigeon shooting must be so built as to meet the rules of the chief clubs: in England the bore must not be larger than 12, nor the gun heavier than 8lbs.'* He advocated pigeon guns that were from ½ to ¾ of a pound heavier than the normal game gun and stressed the need for perfect balance.

Proportion

The typical pigeon gun will be set up, rather like a modern trap gun, to shoot a little high, as birds are on a rising trajectory as they leave the traps. Pigeon guns will strike the viewer immediately as being heavier, more robust versions of familiar styles of shotgun.

Note the dimensions and profiles of these two Purdey sidelocks: on the left is an 1889 game gun of 6¾lbs; on the right is a 1929 live pigeon gun of 8½lbs.

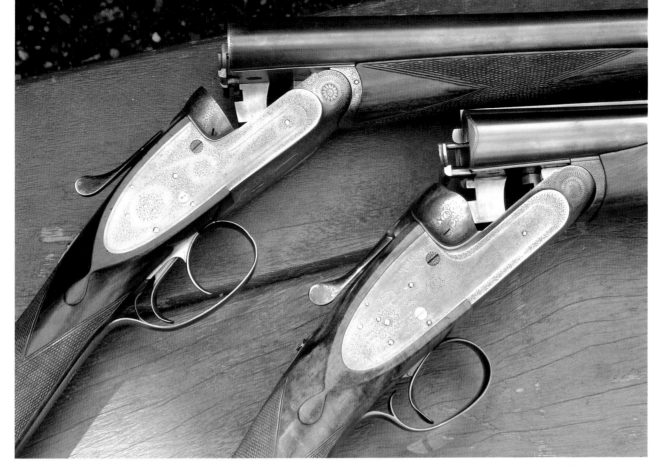

A close-up comparison of the two guns featured opposite. Note the much larger fences and side clips of the pigeon gun and its robust lumps for added strength.

The Stock

The comb will be higher than normal and probably quite straight. A pistol grip or semi-pistol grip is common, as are single triggers on later guns. Sporting percussion muzzle-loaders normally had a ramrod secured under the barrels for loading in the field. In pigeon guns this was often omitted, as gun club rules usually required guns to be loaded from communal supplies of powder and shot and a club ramrod was therefore provided for the purpose. Early percussion pigeon guns of single barrel configuration (and often large bore) will be encountered dating from the 1860s and earlier. Some guns of 10-bore were made for use abroad for a considerable time after the British clubs had restricted users to 12-bore or less.

Barrels

Barrels will be 30" or longer with tight choke boring in both barrels (after 1875) and chambered for 1¼oz of shot or 2¾" cartridges, marking dependent on the age of the gun. Greener again offers a useful observation on choke: *'the largest possible killing circle at one yard behind the trap with the first barrel and at five yards with the second.'* The barrels will appear heavy to the user of a game gun. The rib will generally be raised, broad and flat with a filed or engine-turned surface to reduce glare.

Actions

Actions on breech-loaders will often feature an additional locking measure in addition to the standard Purdey bolt of 1863. The Greener 'treble wedge fast' cross-bolt is a popular feature (after 1873), as is the Scott variant, with a square, rather than a round section, locking bolt. Many pigeon guns will have side clips fitted to the fences to prevent lateral movement under the heavy charges fired.

Actions will be heavy in appearance but will be varied in type with Anson & Deeley and Greener box-locks most frequently encountered, as well as top-quality side-locks and hammer guns in both bar-action and back-action form. Engraving is often distinctive: the appearance of a pigeon on the rib, top-lever or lock plates is a bit of a giveaway.

The Vena Contracta

A friend of mine took his Trulock & Harriss boxlock ejector to a London auction house a few years ago with the intention of getting a valuation. The junior specialist had a look at the action, stock and barrels, declared it what it was – a mid-grade, Birmingham-made, BLE, and then got out his choke gauge and bore measuring equipment: 12-bore at the breech, 2½" chambers and...! Suddenly the junior specialist looked a little agitated.

The barrels were too tight at the muzzles to use the choke gauges. In fact, the barrels were as those of a 20-bore except at the chamber and around six inches in front of it. Burrard describes such guns as 'a 12-bore chamber with a 20-bore barrel [connected] by means of a chamber cone about 6 inches long'.

This is the Vena Contracta. A number of gunmakers sold this style of gun; introduced by Joseph Lang in 1893 (but actually patented by H.F Phillips and built by Webley), to be a 20-bore that could take 12-bore cartridges and eliminate the need to have both gauge shells in the house. The idea was novel, it allowed barrels to be made lighter, and according to accounts written at the time, the gun patterned well and functioned without delivering excessive recoil.

I have seen two Vena Contracta guns and in correspondence have been interested to note that many people confuse the Vena Contracta with the 'Twelve Twenty'. There is no reason for this confusion, as the 'Twelve Twenty' is the name by which Lancaster retailed the Baker back-action of 1906 as a lightweight 12-bore with conventional barrels. The Vena Contracta is the only design that involves the scaling down of the tube dimensions as described above.

My friend also once showed his Vena Contracta to the late Geoffrey Boothroyd, who tactfully commented; 'It is very interesting but I don't know that I should like to have one'.

Quality

Pigeon guns vary in quality, as do game guns but many very finely finished pigeon guns appear regularly on the market. Greener, Purdey, Holland & Holland, W&C Scott and Churchill all specialised in pigeon guns. The reports in the press acted as free advertising for the gunmakers' wares, as they would state not only the name of the winner but also the gun used to win the prize.

Value

Pigeon guns at one time lagged behind best game guns in their realized prices, but they have since gained popularity with collectors, shooters of sporting clays and wildfowlers as well as pigeon decoying enthusiasts. Renowned pigeon shot Will Garfitt shoots with a Beesley boxlock pigeon gun, of which he has a pair and Mike Yardley used a W&C Scott sidelock pigeon gun to win the 2004 Essex County Championship at sporting clays. They are also gaining something of a cult following among shooters of today's 'tall' pheasants, as their tight chokes and ability to handle heavy loads gives them something of an advantage over the more usual game gun.

This W&C Scott 'Monte Carlo B' is a sidelock non-ejector 12-bore designed for live-pigeon competitions. It is a very nicely made bar-action sidelock, introduced in 1891. A new one cost $130 in the USA in 1893. Scott's were better known in the United States as a 'name' than they were in the UK, where they were primarily 'makers to the trade'.

The project (or buying junk for fun)

It is nice to buy an old gun by a top maker in excellent condition and look after it, show it off and maintain it as the fine tool that it is, but there may come a time when you get the opportunity to buy something in very poor condition and wonder what the chances are of getting it back to life. This is 'the project':

The idea will not be new to many of you. People buy old cars and fix them up, old motorcycles too. My mother learned to re-upholster furniture and still sits on much of her work today. However, guns can be seen as a step too far because of the varied nature of their designs, the impenetrable reputation of the gun trade and genuine worries about safety.

None of these reasons should put you off if you have the inclination. I shall consider a project of my own as an example.

The journey (for that is what any project will become) started with an advertisement on the website **guntrader.co.uk**. A gunsmith in Stafford (Bate's) had some tatty old guns brought in by a widow with the instruction to get rid of them ASAP.

Some were in bits, others shootable and others apparently ready for the scrap heap. I bought a shootable C.E. Smith hammer gun for a friend to sell in the States and the Adams & Co with sleeved barrels I have written about earlier in this book. This became a minor project but was usable as it stood. Almost as an afterthought I paid $100 for a hammer gun by J. Thompson because I liked the locks and action.

As a project it looked hopeless at first (explaining its cheapness). The barrels were heavy, blacked, horribly pitted and filthy. They were also slightly 'off the face' of the action. The wood was dented, scratched, worn and had a big gouge in the middle of the stock. The forend did not match the gun, but fitted well enough to be useable. It was proofed for black powder only and was out of proof due to the horrible condition of the barrels.

So far so bad!

So what was I thinking of? As I mentioned earlier, the original idea was to bin the barrels and stock and keep the action as a curiosity. So, I took it home and took it to bits. My instincts about the quality of the gun had apparently been sound. Under all the filth, the action was very nicely engraved with beautifully executed scrolls and the metal-to-metal and wood-to-metal fit was excellent.

The gun was a rotary under-lever, bar-action hammer gun with a high proportion of its color hardening remaining. The locks inside were beautifully crafted with crisp scears and color hardening visible on the bridles.

The hammers were elegant and original and the long top strap equally elegant and finely engraved. Unfortunately, somebody had lost the forend many moons ago and a replacement loop had been brazed on and the forend from another, later gun, attached to 'make do'.

The barrels seemed heavy for steel so I stripped away all the blue and found good, three-iron Damascus underneath. Next, I cleaned out the barrels and took them to Dave Mitchell, the Purdey gunsmith, my oracle when I need expert advice, to get them measured.

My bet was that, although the barrels were rough, they were thick enough to remove the pitting and prepare for proof. Dave measured the barrels very carefully and found them to be 50 thou thick in most places and even at the bottom of the deepest pits there was 33 thou remaining.

After lapping out the pits, the gun was re-jointed and then submitted to the London Proof House, where it passed nitro proof at 850 BAR for 70mm shells (Standard Nitro Proof). The barrels now measure .722 left and .724

The J. Thompson hammer gun after all the dirt had been cleaned off with spirits and fine wire wool.

The forend iron in the mid-stage of renovation. Here it has been welded and re-shaped but not yet engraved.

right, with three points of choke in the left and eight points of choke in the right.

The work was completed by the end of the season ending 2006, just in time for me to use it on the last day – I killed two pigeons with the first two shots and ended the day with a high driven hen pheasant dead in the air to the last shot of the season.

The following summer, Dave Mitchell re-profiled the odd forend iron and had it engraved to match the action, he filled a space in the action invisibly and re-checkered the hand and forend wood. I raised the dents, papered up the stock, recolored it and re-finished it. I was mindful that the restoration had to be in-keeping with the gun, I did not want it to look obvious or out of character and I'm pleased with the results.

So pleased in fact, that work was put on hold for the whole of the 2006/2007 season, while I used the gun almost weekly for driven shooting, including some high-

pheasant days at Bodnant in North Wales as a guest of the very kind Irish sportsman Paddy Byrne, who owns the shooting there. Loaded a little heavier than usual, I used Game Bore Pure Gold 30g shells with No.6 shot and once on my birds, found they were killed cleanly out to 45 yards with regularity.

I found the Jones under-lever to be no great handicap when shooting driven birds, you just need to work unhurriedly and select your shots and great sport was had with no loss of enjoyment from the slightly slower re-loading cycle. I took my Purdey as a back-up gun but never needed it.

I still have a little sympathetic work to do on the Thompson and am still intent on bringing out the figure of the Damascus, but I'll do it in a style that suits the age and condition of the gun. I still wonder by what accident I came upon such an unlikely gem. Not only was I able to salvage what looked beyond repair but I have uncovered a gun with which I shoot very confidently and (for now at least) prefer to take into the field above all others. Happy days.

There can be no comparison between the satisfaction and sense of achievement one gets from returning an old gun of this quality to the field, where it belongs, and the use of an off-the-shelf tool. I wonder who last used the old Thompson, why it fell into such disrepair and where it lay for so long to become so filthy? I wonder who first bought it, who lost the forend, where it saw service and what company it kept? Most of this I will never know, but I do know that it will once again 'live' and be used for the purpose for which it was intended, and that I am responsible for its resurrection. There is no reason that it cannot be used well for many decades to come.

Then we come to the research – who was J. Thompson? Where did he trade? When did he make this gun? What else did he make? Who were his customers? Where was he apprenticed? What happened to him? There is a lot of interest to be had in a little project such as this one so do not be afraid!

You may not feel ready to tackle a major project of the type just described – but start somewhere! When you are offered a tatty or neglected gun or shooting-related item, consider its potential rather than be put off by its appearance.

I have old gunstocks that have broken and been removed for replacement – they provide comparisons of stock shape, they provide seasoned walnut should you

Back to work. My Thompson with a bag of Hertfordshire partridges.

The ultimate project

Let us imagine you have $20,000 and you want a new gun. You always lusted after a Purdey or Boss but even with a substantial sum such as you have, you are still more than $80,000 short of the price of a new one. You worry that a second-hand gun may not fit quite right. Your mind begins to wander in the direction of a foreign side-by-side or one of the largely foreign-made but 'English finished' side-by-sides currently available for this kind of money. Stop!

Consider this: Go to an auction and look at the 'stock, action and forend' section or look for a gun with a broken or really horrible stock. Get all the nastiness out of your mind and think this to yourself:

- A new stock: The finest workmanship by the best craftsman available in England. Made to measure and you cannot get better: $4,000 + wood.
- New chopper-lump barrels by a barrel-maker used by all the top firms: $8,000
- Money left to spend on the most beautiful vintage action you can find by Boss, Holland & Holland or Purdey: $6,000-$7,000.

For $20,000 you have a made to measure gun of your own specifications built on a beautiful vintage action, probably better than a new one, and the total cost is less than a quarter of the price of a new gun by the same men that made your gun.

I challenge you to spend your $20,000 more wisely on *any* new gun currently offered.

While writing this chapter, I competed in the British side-by-side Championship, where I shot with the well-known gunmaker, Jason Abbot, who had the results of just the type of project described above and was putting it to good use.

He had purchased a 1920 Purdey 12-bore sidelock 'live pigeon' gun with a bad stock and worn-out barrels. He has had the gun sleeved very well and fitted with Teague removable chokes. It has been re-stocked to Jason's specifications in a 'Prince of Wales' style with a nicely engraved steel pistol-grip cap in a beautiful piece of walnut. One could not wish for a nicer gun.

need an odd piece. They also provide an opportunity to practise dent raising or smoothing and oil-finishing on old, worthless gunstocks before you try on a real gun.

I bought another filthy bit of junk that had clearly been at the bottom of someone's cellar for years. It was a boxlock gun with cut-away sections in the action and the wood, presumably done by a gunmaker to show customers how the locks are cocked by the falling of the barrels.

I cleaned it, re-finished the wood, removed the pitting from the barrels, filed smooth the cut surfaces, cleaned out the checkering and now it hangs on the wall as a very attractive curiosity. It also demonstrates the workings of the lock perfectly.

Great fun can be had from such little projects – start small and become more ambitious and you will learn a lot about the guns you encounter and the processes involved in making them.

Gun fitting & vintage guns
by Mike Yardley

Author and shooting instructor Mike Yardley has conducted exhaustive investigations into the science of correct gun fitting and has probably tested more guns than any other living writer.

'Gunfitting can be a complex subject, but many people when buying a new gun want to know if it fits them reasonably well. To do one's best shooting, a gun must naturally point to where one is looking and it must control recoil effectively. If the stock is too short it may bruise you, the hand may hit the trigger guard, the nose will be too close to the thumb of the hand on the grip and you will have to exert too much effort to control the muzzles.

It will also be difficult to point the gun well. If the stock is too long, your mount and swing will be impeded. In a 12- or 20-bore gun of normal weight, I like to see a gap of two finger-widths, about 1½" between the tip of the nose and the base of the thumb when the gun is comfortably mounted.

Length may, of course, be adjusted by temporary or permanent means. When buying or evaluating guns, it is always a good idea to carry a 1", slip-on, butt extender to give you an idea of the suitability and feel of any shorter new gun being considered. On light, small-bores, moreover, stock length may be increased to three finger widths, to make them more controllable.

This is an old trade secret of English fitters. On old, heavy and long barrelled, wild-fowling behemoths, by contrast, stocks may want to be shorter than you would normally have on your 12. As with longer stocks for small bores, this is just because it feels *right*.

A recoil pad can add an 1" to a gun that is a bit short (you may have to make an allowance for flattening the butt sole), a wooden extension can give you much more and looks fine when the work is undertaken well and the grain is nicely matched (there are even experts now who can paint in the grain to match an extension to the main section of the stock). Black ebonite extensions are another possibility (one can get about 1⅛" out of these) and one might combine a short black plastic spacer with a conventional pad as well and win a little extra length without it looking wrong.

Drop is a critical dimension in any shotgun; the most critical in my opinion. Gun stocks are bent down slightly relative to the bore or rib axis. This allows the head to take up a comfortable position with one eye looking along and just above the rib. Drop is usually measured at the nose of the comb and at the heel of the stock relative to an imaginary line extending back along the axis of the rib.

Normal dimensions for a side-by-side are 1½" (at comb) and 2" (at heel). Many older guns are too low in the comb (something that may be temporarily remedied by a rubber comb raiser and some vinyl electrician's tape and permanently altered by means of heat; infra-red, hot oil or steam, and bending).

To see if a gun is too low (or too high) for you, mount the proven empty gun into a wall or sky, closing your eyes as the stock comes up to face and shoulder. Maintaining normal cheek pressure, open your eyes. Can you see any rib? If you are staring into the breech the gun is too low. If the rib appears to be climbing *when your head is properly positioned on the comb* it is probably too high. Generally speaking, side-by-sides want to be a little higher in the comb than over-and-unders, they are more prone to flexing in barrels and grip and hence have a tendency to shoot slightly lower.

One may compensate for this by raising the comb $^1/_{16}$" to ⅛" at heel (taking an 1½" and 2" as the standard for side-by-sides, the standard for an over-and-under is 1⅛" and 2⅛"). Long, heavy, side-by-sides may have a tendency to shoot lower than shorter barrelled, lighter, guns.

Cast, bending the stock to right or left to suit your handedness or eye-dominance is not a subject that can be fully considered here. It is the most complex of the variables of gunfit. Suffice to say first that most, but not all, men should shoot off the shoulder that corresponds to their handedness with both eyes open and a bit of cast (*cast-off* to the right for a right-hander, *cast-on* to the left for a left-hander).

A significant proportion (about 30%) of the male population will not be able to shoot with both eyes open effectively because they are either *cross-dominant* or have an eye opposite the rib that causes a pull to one side; they are not pointing the gun where they think at distance. They should shut or dim the eye as they mount the gun. For this group of 'monopeans' the only function of cast is to get the eye looking straight along the rib and the butt comfortably positioned at the shoulder.

For the two-eyed shooting population, though, cast may be used to manipulate point of impact right and left. Someone who is right-handed and absolutely dominant in the right eye will only need enough cast to keep them looking down the rib truly (their requirement in this respect is no different to the one eyed shot). Someone who has what I call 'predominant' dominance in one eye (where one eye is clearly more dominant but the other has a significant effect), may need a bit of extra cast to compensate for it.

In middle age, many men develop eye dominance issues and may need a bit of extra cast (typically ¼" or so) if they are to continue shooting both eyes open. Young shots and women rarely have absolute eye dominance in either eye and are often best advised to shut an eye (consequently they will not need extremes of cast).

To test a gun in a shop or sale-room roughly, use the same method advocated for drop; shut the eyes as you bring the stock of a proven empty gun to face and shoulder. When you open them you should be looking along the rib. Experience will quickly tell you if a gun has abnormal cast.

Other points to watch out for include signs of darkening or staining at the grip (the tell-tale of poor quality bending) and combs that have been notably off-set (as is often the case on Churchill guns) and stock combs that have been 'swept' or 'scooped'. Off-set can be roughly determined by looking at the nose of the comb relative to the end of the top tang (the metal extension upon which the safety slides). Sweeping can be subtle (and hence difficult to spot). Hold the open gun up to the light by the barrels and look along the comb; it will be immediately apparent if the comb's surface is unusual.

The function of offset combs and sweeping is similar to that of cast; to bring the head and eye to one side. People with very wide faces may benefit from such modifications; they can also be combined with cast to avoid extremes of cast at heel and toe (and the dangers of increased recoil and poor placement at the shoulder). For those of normal proportions, though, they are generally to be avoided.

Finally, pitch. This can be made very complicated. Simply put, length is not only measured from the middle of the front trigger to the middle of the butt sole, but also to the heel and toe of the butt (or, to be really precise, to the heel bump and toe of the stock). Most shotguns would point down slightly if you placed the butt on an even floor and noted the muzzle position relative to a perpendicular wall. There is a lot of waffle on this subject.

When fitting, I like to see even flesh contact throughout the length of the butt sole as the gun is brought to the shoulder. Those who have well developed shoulders and ladies may well benefit from a reduced toe measurement. I am old fashioned, however, and like the traditional measurements of an old Purdey which typically show a ¼" extra at toe and ½" at heel. A bit of toe, provided it is comfortable, helps to keep one up on the line of higher birds.' – M.Y.

Mike Yardley in front of a huge 'mountain' of cartridge cases at the West London Shooting Ground during a break from regulating his .416 Rigby buffalo rifle.

Preparing guns for re-proof

Before submitting a gun for proof (or more likely re-proof) it should be inspected by a competent gunsmith to ensure it is sound and in a condition to be proved. Many guns not prepared in advance will be rejected before they ever see a proof charge. The proof house will also do this inspection and report but it is usually cheaper and easier to have a local gunsmith do it and take care of any problems likely to result in rejection at the proof house.

The Proof Houses will not, according to Birmingham Proof House publicity, *'advise as to the mechanical condition, quality, or value; [of a gun submitted for proof] but the report should determine:*

1. Whether the gun has been nitro-proved, bears the relative proof marks and is suitable for use with modern cartridges.

2. Whether the bores have been enlarged since the barrels were last proved.

3. Whether the barrels are in good condition generally: that is, free from rust, pitting, dents and bulges.

4. Whether the chambers have been altered to accept a longer case since the gun was last proved.

5. Whether the action is 'off the face' from the barrels, or otherwise loose.

6. Whether the action is Structurally Sound.'

Re-Proof

Your gun was originally submitted for proof by the maker in time for its first sale; in many cases decades ago. Work carried out on the gun subsequently or damage, wear etc may nullify the proof marks. The Birmingham Proof House advises that:

'Re-Proof is required after a gun has been proofed and used, but a change in condition or dimension may cause re-proof to be required. The most common causes of this are:

* Increase in bore size or chamber length that causes the original proof to be invalidated.

* Enlargement of bore diameter in removal of pitting weakens barrels and is the most common cause of barrels becoming out of proof.

* Physical signs of failure or possible failure such as pitting of the barrel, cracking or bulging.

* Replacement of the barrel, re-sleeving, welding brazing or electrolytic deposition of any form including chromium or Nickel plating.

* Fitting a device such as a muzzle brake, if a rifle or a variable choke if a shot gun.

* Use of a propellant or load which produces a pressure in excess of that for which the firearm was designed. Particularly to the use of smokeless cartridges in a gun designed and proofed for black powder only.'

Deactivation

According to the Metropolitan Police web site: *'Deactivated weapons are any firearms which have been converted, in such a manner that they can no longer discharge any shot, bullet or other missile. More importantly, deactivation is intended to be permanent and such firearms should be incapable of being reactivated without specialist tools or skills.*

Deactivation work carried out in the UK since 1st July 1989 will generally have been endorsed by one of the Proof Houses, the weapon proof-marked and a certificate of deactivation issued. To these ends, any weapon, even a prohibited weapon such as a machine gun, can be deactivated. The outcome is that the weapon is no longer a firearm within the meaning of the Firearms Acts, and consequently may be possessed without a firearm or shotgun certificate and may be displayed in the owner's home, rather than be locked in a gun cabinet.

Amateur gunsmiths beware: rendering a gun incapable of use in your own workshop is not considered 'deactivation' by the British authorities. 'Deactivation of a firearm is not something to be undertaken by the layman. There are stringent requirements before a weapon can be proofed as deactivated and such work is best left to a gunsmith. A Registered Firearms Dealer is the best person to speak to if you require a weapon to be deactivated. He can make all the necessary arrangements for you, including deactivation of the weapon and getting it proofed.

Although the above references to proofing and certification do not preclude the possibility that a firearm which has been deactivated in some other way may also have ceased to be a firearm within the meaning of the 1968 Act (as amended), it is important that care is taken when acquiring any firearm which is described as 'deactivated'. You should ensure that you are shown the Proof House mark and certificate issued in respect of any gun deactivated in the UK since 1st July 1989.'

On proofing but not stamping

Some American gun owners have asked me about the possibility of having their American guns proved in England but not stamped with proof marks. Rather, they would prefer a certificate of confirmation of proof with the relevant information.

Their desire is to avoid 'defacing' their American classics with proof marks that are not 'original' to the gun. British owners appear not to see proofing as defacing and a lot of old Damascus barrelled guns are routinely submitted for later nitro-proof. It has no effect on value (except possibly to increase it). If any American gun is offered for sale in the UK, it must be subject to proof regardless of age.

Mr C.W. Harding of the Birmingham Proof House told me: '*The minimum set of proof marks we are prepared to use are the BNP mark with crown above and the Inspector's private view mark. The weapon would then also have to have a certificate*'.

It is therefore possible to send your American gun barrels for proofing but they will leave the proof house with some marks, even if you elect to have others entered onto a certificate instead of stamped on the flats.

Below: Proof laws exist to protect the shooting public. It is better that this kind of catastrophic failure happens in the controlled conditions of the Proof House rather than in your hand.

Amateur gunsmithing

The buyer of vintage guns will in all probability become a 'tinker'; he will, sooner or later get the urge to start taking his guns to pieces, to undertake basic repairs and perform routine maintenance tasks. This is a good sign for there is no better way to get to know your guns than taking them to bits.

Before starting to work on guns with the intention of maintaining them or improving them, first make this your aim: to do them no harm.

This may seem self-evident but a turnscrew in the hands of one acting hastily or without clarity of thought and deed is an implement of destruction. To my consternation (I hesitate to say shame, for we all learn through our mistakes) I have on occasion over-tightened a pin and sheered it off, or burred a nicely cut slot through imprecise use of the turnscrew. Gunsmiths endure long apprenticeships to become the skilled craftsmen that they are and there is no shortcut. If we are to become amateur gunsmiths, let us not be ham-fisted bodgers. Let us think before we act, learn methodically and know our limitations.

For these reasons, I buy old rubbish and use it for practice before I try my skills on a gun of value. Before having the confidence to oil-finish my Webley & Scott, I must have completed four or five old stocks and two on cheaper guns and developed a feel for what I was doing. Each time, I got more efficient, more practised and more confident. My Webley now has a finish that is better than the one with which it left the factory. While I have a long way to go before I approach the results of a top stock finisher, I am proud of what I have learned. I have since done my Holland & Holland, after all the French polish came off during a very wet day's shooting.

Before starting to work on your guns, get the tools for the job. Desmond Mills and Mike Barnes, writing in *Amateur Gunsmithing* in 1986 recommend the following list and I reproduce it here for the reader who may be considering setting up a workshop at home for disassembly and maintenance of vintage guns:

Oil and cloth: A clean piece of cloth, preferably cotton, plus a can of 3-in-1 (or similar) oil. I also use white Vaseline to smear on certain parts after cleaning.

Feathers: Wing and tail feathers of pigeon, partridge or pheasant are ideal for cleaning difficult holes and corners.

Shaped piece of wood: A piece of wood size 7" x ¼" x 1½" and shaped to allow access into the slot on the body of the action for removal of the hammer.

Wooden dowel: A piece of dowel measuring 5" long and thinned down to allow access into the slot on the action body.

Main spring clamp: This is essential for compressing and releasing the springs on a sidelock and can be purchased for under $40.

Turnscrews: A basic set of three gunmakers' turnscrews (screwdrivers) is needed. They should be hollow ground (using an oval file).

Pin punches: Three of various sizes are needed. I suggest ⅛", ³/₃₂" and ¼". They can be bought at your local tool shop for about $2 each.

Hammer: A small flat pein type, weighing about ½lb.

Engineer's vice: A standard workshop model with 4½" jaws, which obviously needs to be mounted on a secure bench.

Vice guards: Made from scraps of aluminium or lead and measuring 2½" wide. They need to be the length of the vice and folded at 90 degrees. If you can't get hold of some scrap pieces, as an alternative DIY measure simply cut open an empty beer can, and remove top and bottom.

Parts box: Often called a stripping box, an empty cigar or cardboard box is ideal for placing each part in as it is removed.

Snap caps: These are needed to check that the firing pins and ejectors work correctly when reassembled.

Spring clamp lever: Made from a piece of scrap mild steel measuring ½" x ⅛" and 6" long. Cut a slot in one end with a hacksaw.

Peter Dyson and Son www.peterdyson.co.uk will provide you with everything you need that you cannot find elsewhere.

Mills and Barnes also offer sound advice on the ideal bench for the amateur gunsmith, should you be fortunate enough to have the space needed to construct one.

The gunsmith's workbench. A model for those with the space to construct a dedicated working area.

Maintaining your vintage guns

A gun of the period covered by this book will be at least sixty years old and may be as much as a hundred and forty years old. That it has made it to such an age is thanks to the owners who preceded the current one. If cared for, these guns should be good for decades to come and the key is good maintenance.

The Bores

The internal bores of vintage guns have seen various changes and may bear some scars. One reason that so many fine old guns were consigned to the scrap heap was the presence of corrosive chemicals in percussion caps. Early percussion caps and those used in cartridge primers used potassium chlorate as an ingredient and left chloride deposits in the bores. Fulminate of mercury was also corrosive. If not thoroughly cleaned, these deposits would quickly form rust and this led to pitting in the barrels, which, if left unattended, would destroy them.

Modern ammunition is much more 'friendly' but it is important to maintain a clean bore to prevent damage and to minimise the progress of any old flaws the barrels may have. Always clean the bores as soon as getting the gun home. First push kitchen roll down the barrels to remove fouling, then 'four-by-two' or patches soaked in oil, rubbing until they appear clean. Then scrub with a phosphor-bronze brush and repeat with patches again until they show clean. Finally use a dry patch to remove excess oil.

External

After a day's shooting, wipe away any mud or blood with a damp cloth and then press a paper kitchen-towel over the metal parts to absorb any moisture. Remember to run this down the sides of the rib. Remove the extractors by releasing the stop pin and pulling them out. Use a pipe cleaner, cotton bud or paper towel to clean out the recess. Then put the barrels, sans forend, on the top of a radiator to toast nicely overnight. When dry, rub with a lightly oiled cloth and avoid touching the metal with your hands when you put the gun away.

Likewise, rub the metal furniture and action with a silicone or oil-impregnated cloth but do not leave visible oil on the metal parts.

Oiling

Some parts of the gun require a light application of oil to reduce wear from friction. These include the bearing surfaces on the lumps, hook and knuckle. Oil can be applied with a feather, as this ensures only a light application in a measured way. Feathers are also excellent for cleaning-out holes and recesses otherwise hard to access.

Wood

Gun oil must not be allowed to come into contact with the wood. If necessary, a little linseed oil can be rubbed into the wood with the palm of the hand. An oil-finished stock requires no other real maintenance.

Oil damage

Many fine guns have been damaged by over-zealous oiling. Typically, an owner thinks of his gun as a bicycle and a squirt of oil in the striker holes is thought to be a good idea. Unfortunately, this oil will soak into the head of the stock, causing it to darken and become spongy and unsound.

After some time, the gun will become loose and eventually it will need re-stocking. Guns require very little oil. If the bores are oily and the gun stored 'muzzles up', oil will run into the striker holes and thence into the wood. Storing your gun 'muzzle down' avoids this damage. In any case, clean out any visible oil from the bores before storing the gun.

For full details of stripping and reassembly of boxlock and sidelock guns, Mills & Barnes' *Amateur Gunsmithing* provides a step-by-step guide that is easy to

follow and works. I shall not attempt such detail here, for while this book is intended to connect the collector and user of old guns to the methods of maintenance and servicing within the bounds of their ability, it is not an instruction manual for the disassembly of shotguns.

Once you have your basic tool kit and the inclination to begin tinkering, try the following:

• *Disassembly of an old gun of no value*

These can be had in poor condition or as 'stock action and forend' at auction, often for a few pounds. Get a hammer gun, a boxlock and a sidelock of indifferent quality or poor condition and systematically take each one apart, lay the parts out in order and re-assemble them, after cleaning. Do not bother with the removal of springs for now, just get used to removing the locks and pins without damaging them, putting them back so that they all match up and lie straight without having any 'extra bits' left over when you have finished.

• *Clean out the checkering*

Do this with a brass wire brush and steel wool. Soak the steel wool in surgical spirit and dab onto the checkering. Rub a little then clean gently with the brass brush, following the line of the checkering. Repeat the process until the dirt is all removed. A surprising amount of dirt will come out of the checkering and it will look and feel much sharper. A stock that looks like it needs re-checkering is often just clogged with palm grease and can be refreshed simply in the manner described above. An alternative, perhaps gentler, method is to use a paint stripper such as Polystrippa and a camel-hair paint brush.

Re-finishing a stock

There is a great deal of satisfaction to be had from turning the dull and dented woodwork of an old but quality gun back into the beautiful thing it once was. If tackled with patience and perseverance, this is a feat within the capabilities of the amateur. With practice, excellent results can be had and no specialist equipment is needed.

The first job is to strip off the old finish. I use surgical spirit and fine wire wool, wiping with kitchen towel to remove the dirt as it is cleaned off, before it has a chance to dry. When the wood is bare and matt in appearance, inspect it for any dents or cracks, scrapes or other signs of wear. If you do not need to radically repair and even-up the wood and want to retain original patina but remove dirt that is hiding the grain, try 'Orange Glo', a branded cleaner based on orange oil which is excellent for removing old dirt without damaging the finish.

Dents can be raised using a damp cloth and a hot iron. Put the cloth on the dent and apply the iron to force the steam into the fibre of the wood. The wood will expand and the dent will be raised. Scratches can be filled with a mixture of wood glue and the dust from sanding the stock. Force the mixture into the scratch or scrape and allow it to dry slightly proud of the wood.

It can then be filed or sanded down to the level required. Dampen the wood and steam off the moisture over a cooker – gas flame or hotplate – to raise the grain. Sand the whole stock with very fine sandpaper until it is perfectly smooth all over. Use increasingly finer grades of paper and then buff with a chamois leather. The quality of the finish will largely depend on the care spent on these early preparatory stages.

If the stock is very light in color, you may want to deepen the color using alkanet root dye – available from gun shops. Simply put some on lint-free cloth and apply evenly, allowing each coat to dry before applying the next. Remember, the end result will be a few shades darker than it looks at this stage, so don't overdo the color.

Next, I apply gunstock oil (variously known as 'slacum' or finishing oil). Every gunmaker had his own recipe and guarded it jealously.

Richard Arnold suggests:

Boiled linseed oil – 16oz

Spirit of turpentine – 5 teaspoons [most stockers now use terebine driers]

Carnauba wax – 1 tablespoonful

Venice turpentine – 2 teaspoonfuls

Mix together and heat until it simmers. Simmer for ten minutes then allow to cool.

However, Napier produces a ready-mixed formula in a neat little box with all the odds and ends to get you started. These kits are available at gun shops and save a lot of mess and smell. Lin-speed is a similar product which produces good results and Tru Oil is also reputedly excellent when used properly. A good grain filler is tung oil mixed 50-50 with white spirit, though I rarely use a grain filler.

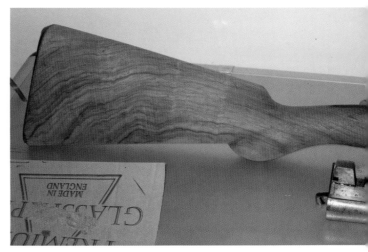

Steel wool and surgical spirit will strip off all the old finish leaving a clean surface for dent raising and sanding to begin.

Apply two coats of this with the palms of the hand sparingly – it will be absorbed. Remove any excess gently with fine wire wool and then leave for two days before you apply a clear grain-filler, allow it to dry thoroughly and then rub it gently with fine wire wool.

Now begin the finish – this must be built up gradually in very thin layers. Rub the finishing oil in with the palm; use *half as much* as the minimum you think necessary. Do not leave any sticky residue on the wood, rub the oil into the wood until you feel it get hot in your hand. Leave a day between each coat and expect the process to take four or five weeks before it is finished. After each coat rub with finishing oil on a cloth; rub hard and then buff with a soft, dry cloth. The results are surprising and pleasing, making all that work worthwhile.

A word of warning: once your friends find out you can do this they will want you to do their guns for them as well as your own. I once got duped into doing the stock of an AYA Yeoman for one of our beaters – he was very pleased with the result but I can't help thinking the effort a little extravagant on the floorboard with which the gun was stocked.

Just by attempting the simple procedures above, you will become far more aware of your guns, how they work and what can be done to keep them in good order.

Finally on this subject, remember that you can damage things more easily than you can repair them, so act with consideration and care. Not everything needs remedial attention. Just think – do you want to lose the

Checkering is a skilled job. Do not attempt it on a gun until you have been properly taught by a gunsmith proficient in the art. Many professional stockers still send the gun out to a specialist for checkering.

patina of age on that gunstock or is it just dirty and in need of renovation? If unsure – try Orange Glo and then polish with Renaissance Microcrystalline Wax Polish, (a best-kept secret of the British Museum) which is excellent for renovating and protecting old gunstock finishes.

Many lovely old guns have been spoilt by amateurs trying to improve them or 'renovate' them with modern ideas of what looks nice. For example, gun actions and lock plates are often polished to brighten them: **don't do it, they were not intended to be shiny!** Let the metal age gracefully. Don't varnish the wood, it needs an oil finish. Don't shorten the barrels or screw on an ill-fitting butt extension. Do not attempt to put checkering onto wood until you are very confident of your skills – practise first! In short, do not embarrass yourself or your gun with unsympathetic bodgery; for some day another will own it and you will want its future condition to reflect well on you.

Cleaning Locks

If you have heavily gummed-up locks, remove them, place in a pyrex dish and pour boiling water over them. Quickly decant this then spray with an aerosol solvent spray used for oven cleaning. Leave for 30 minutes for the oil to dissolve and then scrub under a hot tap with a tooth-brush. Dry the locks with kitchen towel and place them on a warm radiator for 30 minutes then spray with WD40. Clean off excess with tissue paper and oil the locks with a good lubrication oil before replacing them.

Professional Help

If you have old guns, you will need a few good contacts in the trade to do what you cannot, to advise you from time to time and to teach you. Time spent with those who know their trade is learning time. Ask questions, watch, listen and you will surely learn.

'Birmingham Anonymous'

Q. When is a London gun not a London gun?
A. When it is made in Birmingham.

Many fine guns with London addresses engraved on the ribs or actions are living examples of how the Birmingham Trade produced guns in all grades for London makers to pass on to customers *as their own*. The cachet of a London address would attract a premium and buyers who thought they were getting a 'London gun' would pay for the privilege. Some makers even sent barrels from Birmingham to be proofed at the London Proof House with the associated London proof marks adding to the deception.

In some cases, Birmingham firms such as Westley Richards set up outlets in London and put the London address on their guns, though they were made in Birmingham. In other cases London firms like William Evans or Holland & Holland bought the guns from Birmingham and added the engraving, or other 'finishing' to various degrees before selling them as their own work.

Some firms would buy-in cheaper grades of gun from Birmingham to sell under their own name, while making their 'best' guns in London. Purdey and Holland & Holland both did so for a time: Purdey with their 'D' Grade boxlocks and Holland & Holland with their 'Climax' and 'Dominion' back-action sidelocks.

Many a proud owner of a 'best London sidelock' or 'quality boxlock' would be horrified to discover that his pride and joy was partly, or wholly, made in Birmingham. However, in many cases the quality of the work speaks for itself and stands the equal of any equivalent gun produced by the London Trade. Those long-dead Birmingham craftsmen remain anonymous but in their day their best work was as good as anyone's.

London gunmakers and (in brackets) the Birmingham firms who made some of the guns carrying their names.

William Evans (Webley & Scott)
James Woodward (Webley & Scott)
Harrod's (Webley & Scott)
Holland & Holland (W&C Scott/Walter Bennet/C.Bickley)
Churchill (William Baker / A.A Brown)
W.J. Jeffery (Daniel Leonard)
Coggswell & Harrison (W&C Scott)
Charles Lancaster (William Baker)
Army & Navy CSL (Webley & Scott)

These are just some of the London firms whose names I have seen on guns made in Birmingham. But London makers were not the only ones involved in 'badge engineering', as it used to be called in the days of the British motorcycle industry.

Provincial makers did make their own guns; E.C. Green of Cheltenham, for example, made plenty of high-grade guns in his own workshops, as did Gibbs of Bristol. However, many other firms would buy finished guns, engraved with their name and address, direct from Birmingham – or would buy guns 'in the white' and finish them locally.

Most guns with obscure and forgotten names and provincial addresses on the locks and barrels will be what have become known as 'Birmingham Trade Guns'. These cheap guns were sold by ironmongers and small gun shops all over the UK. Furthermore, well-known provincial firms with a strong local following also widely adopted the practice of buying guns from Birmingham.

Makers from various provincial locations retailed guns 'badged' as their own but actually made by Birmingham makers:

Robert Lisle, Derby (William Baker)
Walter Locke, Calcutta (Daniel Leonard)
W.R. Pape, Newcastle (Charles Osbourne)
J. Graham, Inverness (Skimin & Wood)
Hardy Bros, Alnwick (Holloway & Naughton)
Alex Martin, Glasgow (A.A. Brown)

Practices change little with time it would seem, for today one can buy a William Powell 'Heritage Consort', which is to all intents and purposes an English gun in presentation and appearance. However, it is made abroad and finished in Birmingham in the English style.

There is a certain irony in a Birmingham firm buying-in guns to brand as their own, when for the best part of two centuries Birmingham firms were the anonymous manufacturers to the greater part of the British gun trade.

David Becker finishing a stock. David is one of a number of ex-Purdey gunsmiths currently operating as outworkers to the gun trade.

A file, a pair of hands and a trained eye. All combined in the preparation of this 2004 Purdey hammer gun.

A fact of life: gunsmiths take ages to get anything done!

You just have to get used to it. A good gunsmith will have lots of work in hand. Working on guns is not a quick, easy operation. Parts need to be made, ideas tried, adjustments adjusted and good work made better. A good man will do good work and will not be hurried. He will not let work pass until he is satisfied that it is ready and right. This is worth waiting for. You do not want the quickest work, you want the job done properly, frustrating though it may be.

I once broke an ejector guide on my Purdey on the first day of the season and got it back on the last day. No matter, I got acquainted with my new hammer gun that season and no harm it did me. When the Purdey came back, the job was perfect.

If you find a good gunsmith, you are lucky. Build a relationship with him and give him the respect he deserves, for he will have forgotten more than you know (or will ever learn). I have had to force money into the hand of my gunsmith; he would not charge me for regulating an ejector that he had fitted with new springs – because it was a point of pride that the thing should work, having been fixed once.

Remember, these men are often working out how an obscure mechanism works before tackling it. Many of the systems they encounter are complex, imperfect, transitional, over 100 years old and many have been interfered with and damaged by previous owners, meddlers etc. Fixing old guns and keeping them serviceable is not like changing the oil in a car engine and fitting a new set of spark plugs. In most cases, each part will have to be hand-made from a new piece of metal.

The more you get drawn in to the gun world, the more contacts you will make. Pay attention. Keep references in your head and on paper. The gun trade always relied on 'outworkers' to meet its demands and it still does. You can find men who make barrels for Purdey, and those who stock guns for Asprey and Boss and they will do it for you. But you have to find them and remember them for that time when you need their services.

Guide prices for work commonly carried out on vintage guns

It is useful to have an idea of the likely cost of having work carried out – especially when budgeting a potential purchase

Services	Notes	Prices from
Strip & clean action		$170
New springs		$90
New firing pins		$90
Adjustment of trigger pull		$170
Re-jointing		$280
Brush & blue metalwork	Including full strip & clean	$260
Re-blacking barrels		$240
Re-browning barrels		$260
Raise dents and lap barrel		$170
Recess choke barrels		$160
Reduce chokes	Pair	$160
Regulate choke	Pair	$160
Pit removal	Per tube	$150
Strip & re-lay ribs	Including re-blacking	$550
Sleeving barrels		$1,700
New barrels	From	$8,000
Re-checker & re-finish stock & forend		$550
Fitting rubber recoil pad		$150
Fitting ebonite extension		$290
Stock bending		$200
Re-stocking boxlock	Including wood	$1,600+
Re-stocking sidelock	Including wood	$3,000+
Lapping bores	Per thousandth of an inch	$17
Relieve chamber cones		$60
Tighten action		$110

in need of renovation or repair. The following prices are intended for guidance only but should be reliable indicators of current prices offered by UK gunsmiths working as 'outworkers'. Approaching a famous gunmaking firm for the same services will incur significantly higher costs. Bear in mind that starting one job often uncovers further problems. These can add significant costs to the final job.

Learning their trade

Gunmakers go through long apprenticeships to gain the skills and knowledge they need to ply their trade. It makes an interesting comparison to consider the apprenticeship

These contemporary pictures from Purdey show craftsmen at work employing traditional skills. These skills are time-consuming to acquire and in short supply.

as it has developed over the years. By understanding the lives of the men who made and maintain our guns, we appreciate them all the more.

The following accounts are of two gunmakers who have touched my life in different ways and who are linked in other ways besides, though many decades separate their lives. The men concerned are James Purdey 'The First' and Dave Mitchell.

Having owned a mongrel called 'Purdey' as a ten year old, I realized a childhood dream when I bought myself a Purdey gun at the age of 33, so the name 'Purdey' has significance for me that goes back as far as I can remember. James Purdey founded the famous gunmaking dynasty in 1814. Dave Mitchell spent many years making guns for the original Purdey's successors and now keeps mine in good order.

TRADITIONAL APPRENTICESHIPS
How the masters learned their trade

The first gunmaking James Purdey was born in Whitechapel in 1784 and was indentured to Thomas Keck Hutchinson, his brother-in-law, in 1798 at the age of fourteen.

Apprenticeships in those days were tough propositions for a boy. The apprentice basically signed his next seven years over to his master. He had to live with his master, who would feed and clothe him, and answer to him in all matters. The apprentice was forbidden to have sexual relationships or marry during the period of indenture. Drinking in pubs was forbidden, as was going to the theatre and gambling. The apprentice was paid nothing in cash and must have needed great fortitude to endure such austerity. Surely such an exercise in deferred gratification would be beyond the comprehension of modern teenagers.

In return, the master would teach the apprentice all aspects of the craft of making sporting guns. Purdey would have learned how to make barrels, file actions and furniture and make springs, carve stocks and fit wood to metal.

Upon finishing his apprenticeship, Purdey went to work as a gunsmith with the greatest gunmaker of his day, Joe Manton. Under Manton's watchful and masterful eye, Purdey would have perfected his skills and clearly did so because he rose to the position of head stocker at Manton's before leaving in 1808 to work for Forsyth as a lock-filer and stocker. By 1812, Purdey was a freeman of the City of London and he started his own firm in 1814. The rest is history.

Gunmaking in the 1800s was a series of inter-related specializations and entrants to all aspects of the trade served long apprenticeships, typically of seven years. The Wolverhampton gun-lock manufacturers, for example, split their small workforces into forgers, filers, pin makers, and spring makers. In many cases, sons would be apprenticed to their fathers to continue the family tradition. Apprenticeships provided the environment for the passing on of hard-learned skills and meticulous quality control. They were the essential ingredients that ensured the continuity of the exacting standards of manufacture for which the British trade has always been revered.

THE 'OUT-WORKER' SYSTEM
Locks

Look on the inside of the locks of a vintage gun of quality. There will often be a name stamped on the inside lock plate. The name 'Brazier', sometimes with 'Ashes' following on is a common one. Another famous brand is 'Stanton', another 'Chilton'.

The casual observer could be forgiven for thinking that these impressions were the names of workers in the factory of the famous maker whose name graces the *outside* of the lock-plates. Not so, though some firms followed the practice of having craftsmen stamp their initials on the work they did – hence the 'AH' on my Purdey for Alf Harvey the barrel maker. Others, such as Holland & Holland, generally did not, though 'JR' stamped under the top strap will indicate the work of John Robertson, who made most of Holland's 'best' guns before the firm opened its own factory in 1897.

Relatively few makers had every component of their guns made 'in-house'. Even those with large factories often used the out-worker system to increase capacity when the need arose, or to commission work of a particular type or quality.

The 'Brazier' and 'Ashes' (Ashes was the name of Brazier's works address) or other stamps inside lock-plates are actually the marks of the manufacturer of the lock plates.

Joseph Brazier (and other family firms such as Chilton, Dodd, Law, Homer, Stanton and Stilliard) produced large quantities of these components and they were supplied to the Birmingham and London trades and used by many of the great firms.

Most of the firms providing these specialist components, including locks and Damascus barrels, were based in the 'Midlands', around Wolverhampton, but London-based craftsmen were regularly employed for specific jobs as well. This gave gunmakers the required flexibility to expand or contract their production according to demand, while keeping standard overhead expenditure to a minimum. The gun-lock making industry in the Midlands went into terminal decline after 1900, having had its heyday in the years 1860-1890.

Engraving

One thinks of engraving as being stylistically typical of certain makers, especially before the Second World War. We hear of Purdey 'rose & scroll', Holland & Holland 'deep foliate scroll', etc, and each gunmaker did indeed have a 'house style'. However, engraving was often commissioned by the firm and often sent out to self-employed engravers for the engraving to be added according to the demands of the commission, or the price available. Purdey, for example, had many engravers 'in–house' (James Lucas being one renowned for his introduction of the distinctive fine rose and scroll used by Purdey from the 1870s), but there are numerous records of work being sent to out-workers for engraving. Engraving was not traditionally signed and to this day the names of the executors of some of the most beautiful artwork found on gunlocks remain unknown.

Engravers have become 'known' relatively recently but much of their early work was unrecorded and unsigned. Engraving was just part of the job – like polishing the stock. Even greats like Harry Kell (1881-1958) produced anonymous (though recognizable to the tutored eye) engraving to order for Churchill, Rigby, Watson, Purdey, Holland & Holland and others, before eventually moving to work at Purdey's factory in his later years. Saunders and Sumner were other masters of the graver whose work went unrecorded for many years (Sumner engraved almost all Boss guns for a period with his fuller-coverage rose and scroll). Lucas was unknown to most because he worked only for Purdey, as an employee, rather than as an out-worker. He is recorded in the dimension books as 'JL' or simply 'L'.

Barrels

The out-worker system is still very much in evidence today. One may go to a top London gunmaker with a vintage sidelock made by the firm a century ago to have it re-barrelled, expecting to have the cachet of the maker's name on the new rib and the confidence that the job has been properly done by the right firm – i.e. by the makers. However, there is a good chance that the job will be passed-on to an out-worker (albeit a top-class one such as the Holland & Holland trained Bill Blacker).

The gun will come back with 'Famous London Gunmaker' engraved on the rib, as on the lock plates and all will be well and good, the owner none the wiser. It is interesting to note that one could take the gun direct to Blacker and have the job done for around *half* of what the London firm will charge for exactly the same barrels. The only difference is that Blacker will not put a name on the rib to match the barrels unless commissioned by the firm.

Stocks

Likewise, it is entirely possible that your new gun, by whatever top firm you commission it from, will be stocked in the Essex countryside by a man working from a little workshop behind his farm, rather than in the firm's London factory. I have seen such men as Purdey-trained David

Turning a pile of metal forgings into a supremely elegant pair of guns requires rare skills that are learned the hard way.

Becker with guns on the bench for three 'best' London gunmakers simultaneously. This may shock some, but it should not be shocking for it has always been so. John Robertson, later to become the owner of the firm of Boss, was a well-established 'outworker' to the London trade in the 1880s and 1890s, employing several gunmakers himself and providing the actioning and stocking for most of Holland & Holland's best grade guns at the time, as well as for other firms. Incidentally, Robertson is named in patents for both the 'Boss over-and-under' of 1909 and for the Holland & Holland 'Royal' of 1883.

However, it does raise the rather obvious question: why not cut out the middleman and go straight to the craftsmen for your re-stocking and re-barrelling requirements and save a few thousand pounds? Unfortunately, the market expects a gun to bear the same name on the rib as it does on the lock-plates and if it does not, it will be valued lower. Some careful comparisons and calculations are essential before deciding on a course of action.

A 20ᵗʰ century Purdey apprenticeship

David Mitchell currently operates a busy and successful gunsmithing business from his home in Ickenham within easy reach of the shooting grounds of EJ Churchill, Holland & Holland and the West London Shooting School. It is also a short trip into London, where much of the gun trade is still active.

As with many out-workers, Dave trained with the top firm of Purdey and worked there for many years before leaving to become his own master. This is his story:

Like many unruly '60s teenagers, Dave wanted to be a footballer and drifted along with this in the back of his mind, not doing very much. Dave was keen on sports in general and had already developed an interest in guns so an observant uncle had a 'word in his ear' and put the idea in his head that he might consider learning how to make them. This idea festered away and grew further when a neighbor, who worked in the Purdey offices at Audley House, arranged an interview for him.

At the time Dave had a Saturday job in Regents Park as an attendant at the boating lake. This became full time when the boats needed cleaning and repairing in the off-season for the princely sum of $19 per week. When the Purdey interview came, it was something of a shock. From

David Mitchell inspecting a Boss sidelock at Holt's. Dave has been making best guns since 1963.

being a vague idea, as things so often are to teenagers, the moment of truth had arrived. Suitably polished and prepared by his mother with strict instructions on manners and what to say and what not to say, Dave found himself in front of Harry Lawrence, factory manager hearing the words 'Start on Monday, you are going to be an engraver'. The interview was short and involved no test of aptitude or particular interest. Purdey had space to train an engraver; Dave Mitchell was there, so an engraver he was to be.

So the 16½-year-old Dave found himself, shortly afterwards, asking his boss at the boating lake if he could leave prematurely and sacrifice the $19 wage he was currently receiving for the $6 per week Purdey were about to pay him – for a 42-hour week!

Fortunately, the manager of the boating lake was far-sighted enough to tell Dave to get on with it and that he wouldn't stand in the way of the big career move. As a result, come Monday morning, 4th November 1963, at ten

past eight sharp, Dave was introduced to Alan Dudley his new 'gaffer' and discovered that he was to be trained as a finisher, not an engraver as previously informed.

Now, Dave, by his own admission was 'a little cocky' and when asked if he could file flat, said he could. As a first job he was given a lump of metal and told to file ¹/₁₆" off it 'flat'. After considerable time and effort, back aching and forearms just beginning to realize what the next 40 years of gunmaking had in store for them, Dave presented Mr Dudley with his product. *'Just as I suspected; that looks like a fucking duck's arsehole'* came the blunt response and the lump of metal flew into the bin to the sound of sniggers from the older apprentices, who had clearly been through this initiation themselves and now had a chance to enjoy watching the new boy squirm.

Slightly less cocky, Dave began to watch and learn from the masters. He cycled every morning from home in St John's Wood to the factory, a converted stable block at Irongate Wharf, Paddington.

He was sent on errands, made tea, fetched sandwiches, took barrels to Kilburn Lane on the bus for

169

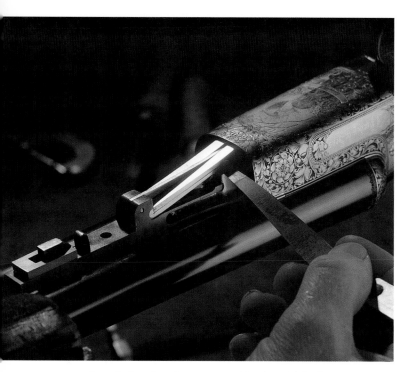

Regulating the ejectors on a Purdey sidelock.

blacking and took them to the Proof House by tube. During all this, he was given metal to file, shown how to stand at the bench. He learned how a file works in one direction only and how it should be wielded. His forearms began to get used to the exercise and build up the strength they needed and he began the process of developing the 'feel' for a file and the eye to judge work by.

First he made a set of turn-screws for himself. But they had to be made the Purdey way. 'Everything had to be done a certain way, in a certain order – everything had to be just right' recalls Dave. Those turn-screws had a mirror finish, polished handles and perfect lines, all by Dave's hand; slow, painstaking, punishing, uncompromising – but 'right'. Nothing else was an option. This was the beginning of the lesson. This was the Purdey system.

These were interesting times socially and gunmaking was not exempt from the turbulence. Older craftsmen arrived at work in suits and changed into overalls at work, starched, detachable collars were still evident and immaculate. These men were Victorians and they were faced with 1960s youth in full cry. The reverberations running through the factory, as through society, can only be imagined by my generation. The Purdey system relied on total control and the old hierarchies and respect for

authority were in decline. Many apprentices did not last long, the regime too unbending, the discipline and exacting standards too much to bear. Other opportunities beckoned and the 'one job for life' convention was on the wane. And remember the wages!

Dave played drums for a band with a regular pub gig in Elephant & Castle one night per week and earned more for that night than Purdey paid him for a 42-hour week – and he got as much free beer as he could drink. He recalls that another Purdey craftsman ran a nightclub.

The grind of learning to use the tools was alleviated with occasional, motivating diversions: trips to the West London Shooting Grounds to help regulate the guns (count the pellets in the pattern, whitewash the pattern plate) and forays into the 'shooting hole'. The shooting hole was a small room at the factory with a hole in the door and a steel plate with a 10 or 20-degree angle to deflect the shot. One simply put a gun through the hole and fired at the plate. All guns had to be tested at various points of development and all those repaired had to be fired before being released to the customer.

As in many contemporary working environments, there was inter-departmental rivalry – a real 'upstairs-downstairs' division in the minds of the personnel. The daily routine was relieved with games of cards; football and music also playing their part. There were arguments and factions and in many respects the environment was secretive, with people keeping information to themselves and even when training apprentices, some skilled men would keep a few secrets to themselves rather than pass them on to potential rivals.

Health and safety was not then what it is now. Those using the shooting hole had no ear defenders and after ten or twenty shots couldn't see anything because of the fumes. Elsewhere, case hardening involved the use of potassium ferrocyanide and when the hardening of 'pins' was going on downstairs the workers *upstairs* would complain about the fumes. Dave recalls that it was hard to breathe where the work was going on and the fumes were so noxious that a naked flame in the room would change color as it came into contact with them. He recalls another craftsman eating sandwiches with hands black with carborundum paste. He also recounts stories of workers in the past baking eggs on coke shovels, though the practice had ceased by his time.

In this environment, Dave continued to learn the art of the gun finisher. Between polishing parts given to him by his 'gaffer', bus trips with parcels of barrels, lunchtime football matches, trips to the shooting hole, lots of filing and sometimes just watching his gaffer do something properly to learn the right way, the only way, the 'Purdey way' to do each and everything that a finisher needed to master.

An example of the fastidiousness of the drip-drip approach to training was the rate at which trainees were given jobs to do. Polishing for example. The parts of the lock would be placed in a 'stripping box' of shavings with a clean box placed next to them. Each part had to be taken out and polished by hand, in order – and the order was always the same.

When each piece had been polished to the satisfaction of the gaffer (when it looked like chrome), it was placed carefully in the clean box and the next part could then be polished. Every polisher did it in the same order, in the same way. Dave's first lock took him a week to polish.

Likewise, the dismantling of a lock (always the right lock first) was fixed and had to be done in the Purdey system, in a certain order, no deviation. As an apprentice learned to dismantle a lock (after his gaffer had removed the mainspring and shown him how to do it) he learned what every piece

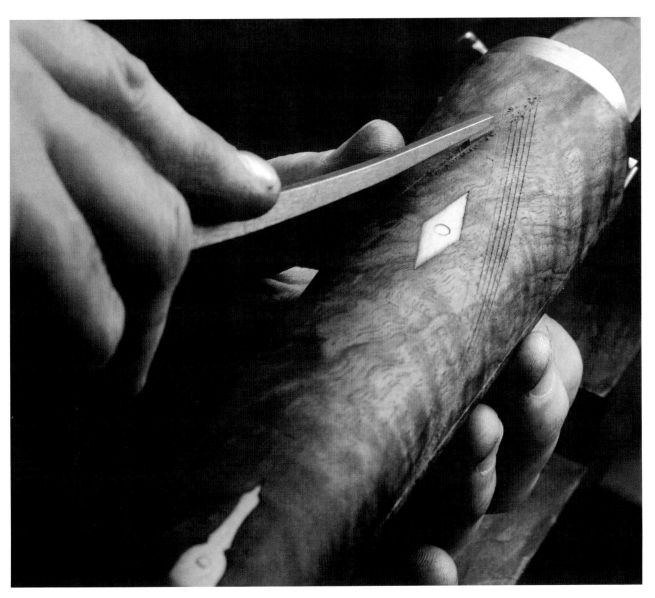

Checkering a Purdey forend. Checkering is an extremely skilled job. Poor checkering will ruin the appearance of all the preceding hard work.

was called and what it was for. *'What's this? Yes, and why is it called a bridle? What does it do?'* etc. When trusted with disassembly and proficient at this, the re-assembly would follow. Dave pointed out the damage an unskilled person could do with a turn-screw and emphasized the gradual building of competence in relation to the rate at which apprentices were given new tasks to master.

Next came the woodwork. Gunstocks would come from the stockers to the finishing shop and the apprentice would sandpaper them with ever-finer grades of paper. Each grade had to be approved by the gaffer before the next could be applied. The system involves each grade of paper removing all traces of the last until a perfectly smooth surface is the result. The challenge is in ensuring the flatness of the sanding surface and retaining the sharpness of the drop points and angles at the butt. Stopping exactly where the checkering began was also an exacting demand. Then, when the final fine-sanding was complete, the gaffer would take out his personal chamois leather and this, applied with spit was rubbed all over the wood until it shone.

On his twenty-first birthday, one change of 'gaffer' later, Dave's apprenticeship ended, under Peter Chapman. Apprenticeships always ended on the twenty-first birthday, regardless of when they began. Dave now began the next ritual – making a new set of tools; a brace, hammers and spring clamps. Stockers would make checkering tools and actioners made jigs and patterns. Barrel makers made strikers and gauges; the tools they would use now they were gunsmiths. He became a member of the Worshipful Company of Gunmakers and a Freeman of the City of London.

Dave emphasizes what he refers to as 'The System' and it plays as a constant chord throughout his narrative. The system was developed by gunsmiths of old, working on a piece-work basis, as the most efficient way of getting a gun through all the stages of production. If everything was done in the same way, in the same order without deviation, then each craftsman could do his job properly without hindering the job of the next man – every stage had to be the same every time. This way guns got produced in optimal time and the craftsmen got paid more. The system worked.

Dave worked in the finishing shop for around five years and then moved to the new lock-making department. The development of this department at the Purdey factory was a sign of the times. Gunlock makers to the trade like Stanton and Chilton began to find lock-making unsustainable and unprofitable and each in turn closed.

Purdey faced a shortage of the locks needed to sustain production. Burt Woolmer was assigned the job of devising the necessary jigs to make lock plates and Dave Mitchell was sent to join him.

At busy times of year, as well as making locks, Dave was put to work on the repair of customers' guns, consequently learning a great deal more about the overall functions and malfunctions of the whole gun – including those not of Purdey make. In time a barrel maker was seconded to the new lock-making department and Dave trained him in the necessary skills and subsequently returned to the finishing shop. This type of movement was not unusual at Purdey as availability of skilled men in each department fluctuated over the years.

The system employed at Purdey for the allocation of pay in relation to work was intrinsic in the functioning of the apprenticeship scheme. A gun-smith may be paid a set wage for a 40-hour week but each job had an expected number of hours allocated to it. For example, to 'finish' a gun may be allocated 70 hours work. If the work was done in 70 hours, then the craftsman was 'on par'; if it took him longer, he would be below his expected level of productivity. When pay rises were offered, they were often set according to productivity to date. However, if he was able to do the job in 60 hours, the 10 hours in hand could be banked.

The strict multi-tier quality control process ensured work was always of the high standard required before it was passed. Each apprentice attached to a skilled man allowed him to count six hours per week as 'time done' so, a skilled man with three apprentices could have his time sheet for the week partially filled-in before he even picked up a file on Monday morning. A man without an apprentice had to do all his own work and only had one pair of hands with which to do it.

Now, if a finisher had an apprentice who could do some of the basic work to the standard required, he was set to do so and the finisher could spend his time on the more skilled work. Two people working on the job got done faster – money in the bank for the skilled man. Thus, it was in the interests of a skilled man to have an apprentice or two – he could use them to make money; but only if he trained them well. The first six to eight months' spent training an apprentice was an investment of time in the boy that paid dividends later on.

Dave explained the intake of apprentices as it occurred in his early years at the factory as typically being between a third and a half of the number of skilled men.

Soot from a dirty flame has long been used by gunmakers to achieve a perfect fiitting of parts.

At its peak there were around forty skilled men employed in the factory. Intake of apprentices was in three blocks, according to the age of the boys in each school 'year'; some would arrive in August, another batch at Christmas and the final cohort at Easter

The hardest thing to learn and to teach, according to Dave, is the absolute rigour of the discipline required to adopt the strict Purdey system of gunmaking – and to file flat.

> Bear in mind that the obsession with filing hit you in the finishing shop – it was magnified to the power of ten in the action shop and barrel-making shop where skill with a file occupied a great deal more of the job.

Teaching an apprentice to feel in his hands and see with his eye what is needed to work to the tolerances required is extremely difficult and for some impossible. It is akin to having a musical ear or sense of rhythm – some will never have 'it,' whatever 'it' is.

The rate of attrition was high, many apprentices not making the grade or unable to cope with the culture. Dave trained three of his own apprentices. One is a highly respected Purdey gunsmith today. The other two did not make it; sadly, one decided the life was not for him only months before the end of his term as an apprentice, despite being an excellent prospect.

It is clear that to produce one top gunsmith, the big firms had to take on, invest time and money in, and subsequently lose, a high number of apprentices; a risky and expensive proposition. It is therefore perhaps understandable that the numbers taken on in the 1980s and '90s dropped to almost nothing, as a cost-saving measure. Happily today, Purdey are again training the next generation of best gunsmiths.

The author and Dave Mitchell discussing their favorite subject, in this case a William Evans sidelock.

THE FUTURE IN SAFE HANDS?
The state of the modern apprenticeship

Richard Purdey described to me, in recent correspondence, the current Purdey training programme as follows:

'At Purdey's we currently have five apprentices (though the old type apprenticeships no longer apply as such, so they are more correctly described as trainees) in:

- *stocking (one),*
- *barrel making (two)*
- *and engraving (two).*

This is higher than it has been for some years and amounts to 15% of our trained gunmakers'.

With regard to pay, the old system of apprenticeship has been replaced and 'trainees' are now paid a wage while they train, as Mr Purdey explains:

> *'Trainees can expect to earn about a third of the hourly rate a fully trained man is paid, though their rates are reviewed and increased according to their skills and progress. As an example the trainee stocker, who will complete his five year training period in December, is now earning about two thirds of what he will be getting once qualified. We also have a finisher who is on a fast-track training programme (actually working under his father)'.*

Mr Purdey continues to explain a new development in the endeavours of the gun trade to ensure a supply of skilled craftsmen for the future:

> *'The Gunmakers' Company has recently set up a Charitable Trust which is setting out to raise a fund from which the income will provide 'bursaries' to encourage youngsters to train as gunmakers and older craftsmen to retrain to be able to take advantage of new computer controlled machining technology.*
>
> *As things stand at present only Holland & Holland and ourselves are running fully integrated factories with our own machine shops, computer technology, and with craftsmen engaged in all seven of the traditional gunmaking trades of barrel making, action filing, trigger and lock making, ejector making, engraving, stocking and finishing'.*

Vintage guns and the future

To conclude this chapter, I shall return to Dave Mitchell for his thoughts on the future and the place vintage guns may have in it.

Machinery plays a central role in the gunmaking operations of Purdey, Holland & Holland and Westley Richards and is a point of criticism for many writers. Purists will say that the disappearance of totally hand-built guns indicates a loss of quality in the finished product: that machines detract from the mystique and the aura of a 'best' English gun.

However, Dave tells the story of a top action-filer taking a pair of actions to a CNC machine operator to see if he could reproduce the dimensions of the pair exactly. A great deal of time was taken to explain the tolerances required and the exactitude of the task. When the action filer returned he was shocked to find his dimensions ridiculed by the operator and proved to be far less exacting than he thought – the computer-generated measuring and production tolerances exceeded that of even the most cultured human eye.

Machining has speeded up the donkey work involved in making new shotguns, the filing starts with forgings far closer to the final dimensions than previously possible. The skilled man-hours can be targeted where they are of the most benefit and as a result Dave believes guns being produced today are as good, if not better, than they have been for generations. He is impressed by the immense efforts that have been made in recent years to ensure that Purdey guns are absolutely of the very best. The use of modern technology has played a big part in the realisation of this aim.

However, in his heart there is a place for the masters of old. When asked about his personal favorites, Dave has a confessed bias towards the classic Purdey action, which is understandable, but he also rates the 'good, robust' Holland & Holland Royal, the 'light, dainty' early Boss sidelocks of the late 1800s and the 'very clever' Dickson: 'When you pick up a Dickson round action, you just want to go shooting'.

However, Dave saves the most telling comment for last and it is the correct one with which to finish this chapter. It concerns old hammer guns: 'Looking at some of the locks makes me feel inadequate, the workmanship is beyond belief'. And it was done without the aid of CNC machinery – and by gaslight!

Purdey's in 2005 had 5 trainees learning the seven gunmaking trades. The system has changed from a traditional apprenticeship to a more modern programme of work-based training. However, the skills at the heart of best gunmaking remain the same.

David Sinnerton, one of the most highly regarded gunmakers working today. As well as working on guns for well-known London makers, he has produced guns of exquisite quality bearing his own name.

PART III

USING VINTAGE GUNS

Equipment

Mike Yardley calls this 'impedimenta' and the word perfectly sums up the essence of the encumbering nature of the various bits and pieces that we shooters load ourselves down with.

In this section of the book, eccentricity gives way to practicality and seeks to evaluate the right equipment to accompany the shooter for a lifetime in the field.

Let's face it: men have a propensity for hoarding things that is second only to magpies. If fishermen are the worst culprits, and I believe they are: consider all those floats, plastic lures, hooks and flies and the various other shiny things that your average fisherman acquires but never uses or needs – then shooters are not far behind.

The more varied your interests, the more you will acquire and God help you if you decide to get into black powder and muzzle-loaders!

Some readers will doubtless have a wealth of equipment and need no advice on it from me but I include this section for those newer to the sport, those returning to it, or for younger readers. Others may chose to skip this section now, but why not read on and see how far your thoughts coincide with mine on the subject?

The following is a list of what most shooters will accumulate and find more or less useful:

- A gun slip
- A cartridge bag
- A cartridge belt
- Wellies
- Leather boots
- Turnscrews
- Camouflage netting & poles
- A 'Pigeon Magnet'
- Cleaning equipment
- A torch
- A game bag
- Camouflage clothing
- A shooting suit
- Hats
- Knives

My prevailing attitude to all of the above is this: buy well and buy once. There is little as irritating as using poor quality equipment. Each time you touch it or see it will cause you displeasure and you will end up with lots more poor equipment to replace the first unsatisfactory purchase.

In contrast, the right equipment will add to your pleasure and will age with you in a very satisfying manner. It will eventually save you money. Buy the best you can get and only buy what you will use.

If you are considering buying your son or daughter something on the above list, do not scrimp. Rather than buy a whole set of third-rate gear, select one item from the list and get the best. Certainly with leather goods, they will outlast you. Long after you have shuffled off to the great pigeon hide in the sky your, now adult, child may reflect fondly, as he loads from the cartridge bag you gave him for his 16th birthday.

Cartridge bags

I am very fussy about leather in general and cartridge bags in particular. I have spent a great deal of time scouring game fairs, gun shops and auction houses to see the various offerings of past and present manufacturers and this is an area where you certainly get what you pay for.

If you have a lot of guns it is a good idea to have a lot of cartridge bags. If possible, only use a certain bag for one size of cartridge. I have 12, 16, and 20-bore guns. To further complicate matters I have 12-bores with 2½" chambers and others with 2¾" chambers. None of these mix with a result other than tears.

Different days demand different bags. I like to carry a big bag of 100 capacity and one of 50 capacity to a shoot. If walking, I can fill up the smaller bag as the day goes on and save my shoulder without running short of shells.

On driven days, I like a 100 capacity pigskin 'Payne-Gallwey' style bag with a metal bar running through the top of the flap. This means it can be kept open during the drive and loading is made much easier. This is also true in the pigeon hide.

Pigskin is absolutely the best material for a cartridge bag. It holds its shape and stiffness for well over 100 years. I know; I have one that age that belonged to a much-loved adopted great uncle. Everybody should have a 100 capacity pigskin cartridge bag of the Payne-Gallwey design.

For general use and smaller bags, cowhide is generally used and is excellent if of good quality, miserable if of poor quality. The better bags, and the ones you should go for, have a canvas lining stitched tightly to the inside of

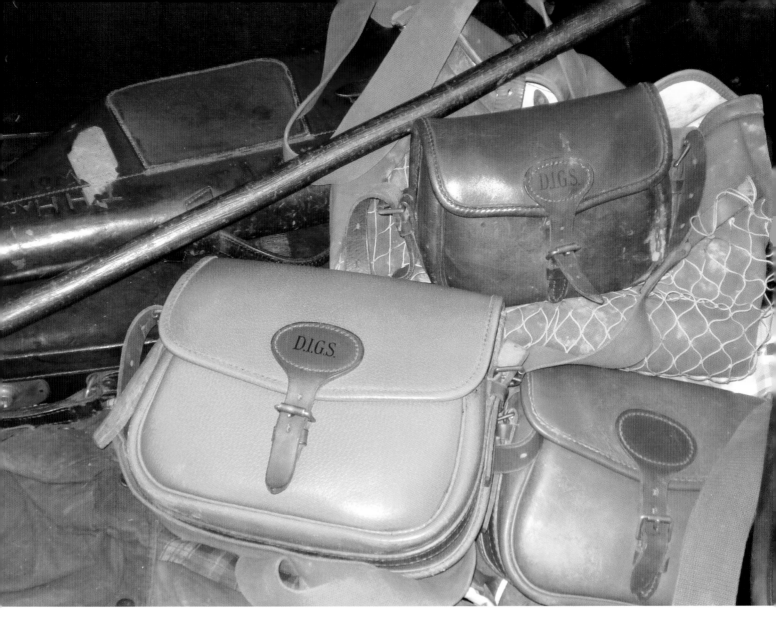

the hide. This helps the bag retain its shape and will make loading easier for the life of the bag. Without this, they tend to sag after a couple of seasons. I have seen this happen to very expensive bags as well as cheap ones.

Anybody can make a bag that looks good in the shop window. The quality only shows after it has spent a few seasons kicking around in the bottom of a gun bus!

In my opinion nobody makes bags to better those by Brady of Halesowen. They are not cheap but they cost less than your girlfriend or wife probably pays for a pair of party shoes and they will last you all your life.

Gun Slips

Many shoots insist on guns being sleeved between drives these days. A gun slip protects your gun from minor

Quality game and cartridge bags will last a lifetime of hard use.

knocks when getting in and out of cars and makes it easier to carry along with other equipment when wildfowling or pigeon shooting. It also prevents you frightening the public at the start and end of each journey, or if roads need to be crossed or followed for a distance.

So, you need one.

The best are very expensive. I have many friends who shoot with guns that cost less than a top-notch gun slip.

They are generally of canvas or leather and most these days are lined, although I have a very old Brady unlined canvas and leather sleeve of my father's that is actually very useful and takes up much less space, being easy to roll up and put in a game bag.

The 'leg-of-mutton' case is unfashionable nowadays but is a discreet and practical way of carrying a shotgun and offers better protection than a slip.

The old 'leg-of-mutton' style of gun case is an unusual sight these days but provides good protection and is unobtrusive; the shape not immediately as recognizable as a gun as is the case with a gun slip. They can occasionally be found in second-hand shops and market stalls and often appear at auction.

Full coach-hide slips are very nice if you have the cash but something of an unnecessary extravagance, a decent quality leather slip or a very good canvas and leather one will do just as well.

Best guns were usually supplied in leather or oak and leather cases. They provide good protection for guns in transit. Good quality vintage examples fetch high prices at auction.

Parsons is a company renowned for the quality they build into all grades of slip they make and, again, Brady comes up trumps with proven designs and tried and tested quality.

I tend to use a sheepskin-lined Brady slip of canvas & leather with a full-length plastic zip and blocked end cap. This has a weight advantage over a full leather slip. The plastic zip ensures your gun will suffer no scratches and being full-length makes the slip easier to dry and clean. Real sheepskin maintains its pile better than synthetic alternatives and gives natural cushioning.

Traditional 'motor cases' were usually supplied with a new 'best' gun and they are nice to own and present a gun well. They provide full protection when travelling with a gun stowed in the boot of a car but can be impractical in a crowded gun bus on shoot day. They are expensive to buy new and vintage examples achieve prices between $400 and $1,600 at auction. For air travel, lightweight aluminium cases with foam padding provide higher levels of protection and withstand rough treatment better. They can be had from around $100.

Cartridge belts

These seem to be going out of fashion. Most of my friends load their pockets from their cartridge bag and I have adopted this habit as well. The only time I use a belt these days is to keep one stocked with bismuth shot, in case a drive is likely to involve ducks. The belt makes choice of cartridge easy and keeps toxic and non-toxic loads separate.

The closed-loop variety is the best and easiest to load from. Go for hide rather than suede or canvas and absolutely avoid the vinyl variety. The belt should be as light as practical but sturdy enough to stand the pull of a full load of 25 shells.

The exception to the 'closed loop is best' rule is when ferreting. While open loops are slower for loading, they are less likely to shed cartridges while one is crawling around on the floor. I have a canvas and leather 20-bore belt with open loops and it is excellent. When ferreting, my pockets tend to be filled with all sorts of rubbish: knives, nets and string etc. I like to have shells easily to hand and the belt provides the answer.

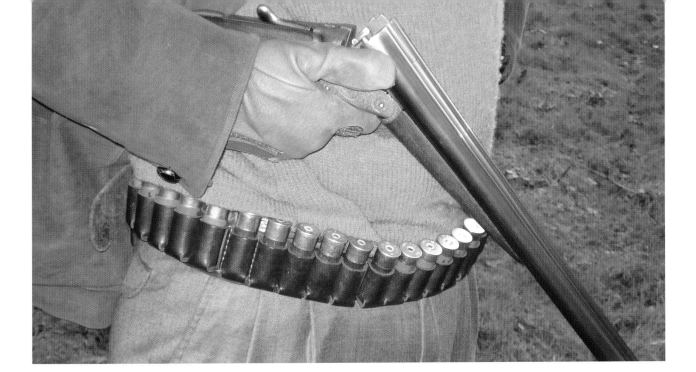

Wellies

These really are a 'must' once the winter sets in and the ground is wet. The variety on offer is vast and they range from under $20 to over $400 in price.

I personally like the neoprene lined type for cold days. They need to have good, grippy soles and be tough enough to plough through brambles without damage. Mine are abused and kept in the boot of my car, under my pigeon magnet, winter and summer.

I am always amused by the stock shooting people in Britain put in clean wellies and have more that once been teased or chided for turning up with dirty ones. If you are planning to make a good impression, make sure they are not strangers to the hose.

Leather boots

You do not want to be walking the fields after summer rabbits in wellies any more than you want to be standing at the peg shooting partridges in them on a sunny September afternoon. Wellies are a necessity on cold, wet December days but otherwise a good pair of leather boots is the order of the day.

A good general idea is to have a sturdy pair for serious walking. These can be of the type ramblers and hill-walkers use, or more traditional styles as retailed by Hoggs and available in gun shops. Whatever you choose, they need to be watertight to the ankle at least, with a sewn-

Leather cartridge belts with closed loops, like this one from William Evans, enable relatively fast loading and spread the weight evenly. The safety benefits are also worthy of consideration. Inadvertently loading a 20-bore shell into a 12-bore from a cartridge belt is very unlikely, while the risk of a 20-bore shell lurking in a pocket or bag is always there to the owner of multiple gauges.

in tongue and sturdy sole with good grip.

I like to have a well-made but light pair of boots as well, for hot days when the ground is hard. The best I ever had were by Camel, but they are discontinued. Canvas and leather combination boots in the safari style are also a good bet, being strong and protective but light and cool.

A shooting suit

Mine is tweed and from Holland & Holland. It consists of a pair of breeks, a long waistcoat with big pockets and leather shoulder facings and a loose covert coat with big pockets for cartridges, buttons to keep the flaps open and side pockets to keep the hands warm.

I bought it in the Harrods sale one year at a 60% discount and do not intend to buy much else for a few years. It is warm on cold days, cool enough when the temperature rises, or walking is required and it keeps me dry in the wet. I tested it the day England won the Rugby World Cup. The foulest weekend on record in my shooting book and one on which I was shooting a 300-bird day in Hertfordshire.

Fashion changes. However, much of our traditional shooting wear endures: a shooting suit circa 1911 from Greener's *The Gun and its Development* (left); John Rutter in 2004 (centre) and a Purdey advertisement from the same year showing a modern take on traditional field clothing (right).

The rain fell like God was using his power shower on us and the wind blew it this way and that. I remained dry and comfortable and what more can one ask? Except to shoot well, which I didn't.

Most of the time I shoot in the waistcoat and breeks only – the coat goes on when the rain comes down and comes off as soon as it stops. I like the freedom of movement and it is warm enough, with a jumper underneath it, all season long.

My girlfriend says I look like Rupert Bear in it; slightly more flattering than A.A. Gill's description of traditional shooting attire as *'dressing up like a Victorian rent boy'*!

Another traditional material going out of fashion is 'oilskin' or thorn-proof oiled cotton, as seen on classic Barbour jackets. The disadvantages are that it is not totally waterproof, it goes stiff in the cold and it shrinks from its original size when it gets wet – so buy one a little bigger then you think you need.

I have a couple of old Barbours that fit all right but the sleeves are now two inches too short. Very annoying. I think the material has a certain charm but I'm not surprised it is being superseded by modern 'technical' fabrics, which are much better in every way (except maybe in resistance to brambles and hedges).

Wearing breeks requires you to wear long woollen

socks. This system is old fashioned but practical. Water will not claw its way up the fabric and you will remain comfortable and dry all day. I buy three pairs of socks from Purdey each year at the Game Fair. This meets my needs for a cost of $50 per year. All well and good. Under the wool, I wear a pair of thin nylon knee-socks, Italian and apparently stylish but like women's stockings really. They prevent the woollen ones becoming itchy or falling down and also prevent rubbing and blisters.

In cold weather, a tight, thin thermal under-layer is a good idea and very effective.

Hats

For game shooting I prefer a traditional woollen tweed cap to keep the sun out of the eyes, the rain off the head and the beaters from laughing at my hair.

Pigeon shooting requires a wide brim camouflage hat. I think normal DPM is as good as anything but get 'Realtree' or 'Advantage Timber' patterns if it makes you happy.

Inland duck-flighting or pigeon shooting in the woods is also best served by a wide brimmed hat but I prefer a tweed one in rusty shades to the usual DPM canvas. Tweed is warmer in the winter and is totally non-reflective.

Camouflage Clothing

I have to admit to being a bit of a traditionalist myself when it comes to attire for a game shoot, whether of the formal driven variety or the less formal 'walk and stand' type. However, there are times when camouflage clothing is essential and those times tend to be the more solitary shooting experiences that we get from pigeon decoying, roost shooting, stalking rabbits or duck flighting.

This is one area where I am not convinced that a lot of money needs to be spent. Army surplus stores are full of different varieties of DPM clothing to suit different terrain for very little outlay. This is also tougher and more tear-resistant than many common brands of 'Realtree' and 'Advantage Timber' clothing, though these are getting better.

Have a set of predominantly khaki and brown patterns for summer stubble fields, a greener set for the woods and hedges and something more autumnal for that time of year. A DPM jacket should cost less than $40,

The hands can be a giveaway – even when the rest of you is well camouflaged.

trousers are about the same and long sleeved shirts can be had for $10 (ideal for a hot pigeon hide).

An essential part of the art of camouflage is to make sure there is cover behind you so that you do not make a silhouette against the sky – and to stand still. Too many people move around all the time and then wonder why the birds don't come in. Pigeons have very good eyesight and they don't want you to shoot them!

I also think face-masks are under-valued as essential parts of camouflage. There is no point being clad from head to toe in 'Realtree' if your big shiny white face is staring up at the sky for every pigeon or duck within 300 yards to see.

There is a standing joke where I shoot that when we do a duck drive off the gravel pit to start the day, I am always in the 'lucky spot', whatever peg I draw. I think it is no coincidence that I am the only one to cover my face before the drive starts and I make sure that I stand still.

For this purpose, I use a scrim mesh scarf from

The author pigeon shooting in Oxfordshire with a 1904 Edwinson Green and a 1920s Hill, both cheaply bought and recently restored.

army surplus, $6 and easy to carry in the pocket.

The hands also act as a give-away, being whiter and prone to movement; so a pair of gloves is a good idea. There are camouflage patterns available in a variety of materials but I prefer to wear a leather glove on the left hand to protect it from hot barrels. I prefer not to wear a glove on my right, I lose the 'feel' I want when I do so. If I do wear one, it has had the trigger-finger removed to the knuckle. Those who can shoot in gloves should do so when camouflage is an issue.

Some shooters cover the barrels of their guns with camouflage tape in a non-reflective material. This is sound thinking but for the user of vintage guns, not something that may appeal. However, you may consider it if you have an old gun you keep for the sole purpose of pigeon shooting or wildfowling. You can always take it off.

I don't camouflage the gun. I build my hide high enough at the front to ensure that I can keep the gun out of sight until the time to shoot arrives. This is a compromise I am happy with but it is important to keep the gun out of the glare of the sun, as it will shine bright and be a complete give-away, signalling your presence from a great distance

Cleaning equipment

Richard Arnold wrote in *Pigeon Shooting*: '*Let the sportsman remember that it is not the use of a gun which shortens its life, but how it is either cared for or neglected when not in use*'. This is worth committing to memory, for it is true indeed.

Cleaning rods – get a lot of them and have them standing in an umbrella stand or something similar. I use my father's old Regimental drum. You need three in each bore of gun you own: one for a jag, one for a phosphor bronze brush and one for a wool mop. This avoids having to mess around with oily threads and makes cleaning much quicker.

I like the heavy type of rod in hardwood with brass fittings and a slightly bulbous end for a good grip.

You will need kitchen roll to clean the initial dirt out of the bores, patches to clean them further and plenty of oil: Rangoon or Parker Hale Express or something similar. I buy the latter two items in bulk for hardly any money and find this a boon – you don't want to have to scrimp on oil and patches.

'Four-by-two' cloth is better, I find, for cleaning dirt from the bore than normal patches. Soak it liberally with gun oil and push it down the bore after phosphor bronze brushing. You can cut it to the size you want for a good fit and it is cheap. I buy six rolls at a time.

When cleaning out a gun on the floor of the gun-room or study, or wherever you have to be, you inevitably find yourself faced with the question of where to rest the muzzles while pushing the rods down the bore. I avoid damage to floorboards by placing a clay shooter's leather shoe protector on the floor and resting the muzzles on this.

Knives

I realize that this an area that can become obsessive and lead to a collection habit all of its own. Most shooters buy knives almost as fast as they lose them, so little needs to be written on the subject save some practical advice on some useful types and good sources of replacements.

For general use, the Opinel range is excellent because the steel is high carbon and retains its edge and the blade shape is good. The locking device and handles are strong and they are cheap, so you won't lose sleep when they disappear after a day's ferreting. Sizes 6 and 7 are the most practical in the range and there is also an excellent folding saw-blade version.

For pigeon hide building or ferreting I like to carry

The 'Blackie Collins' kukri-inspired camp knife: ideal for the pigeon shooter.

a Blackie Collins designed 'kukri' shaped camp knife with a blade big and heavy enough to chop with, yet sharp enough to cut like a knife and light enough to wear on your belt.

An excellent source of knives is the British Knife Collectors' Guild. They have an excellent, informative website and a yearly-published catalogue.

A torch

There is always a time when you will be out later than you planned or find that although the sun was bright and everything looked easy when you set up for a duck flight, finding your way back looks infinitely more complicated in the dark. The answer is to keep a small, (AA size) powerful, torch in your game or cartridge bag.

Maglite torches are light, have beams adjustable to sharp focus or wide angle and they are very robust. I once lent mine to a friend seeking a Canada goose that was dead on the water but which his dog was reluctant to retrieve, after an evening flight on a flooded gravel pit.

He retrieved the goose but dropped my torch in the water. He found it a month later in the mud, where the water had receded, and it switched on first time. I still have that torch and it still looks and works like new – even after spending the whole of November under water.

Torches are available in numerous colors. 'Silver' is the easiest to spot when dropped and, as it is actually polished aluminium, will not wear off with time.

The 'pigeon magnet' and decoys

The 'pigeon magnet', as rotary decoy devices have become universally known, is a marvellously eccentric development in the pursuit of the pigeon and represents a high point in ingenuity that can produce excellent results if used properly in the right conditions.

Original machines are simple in design but have become more robust, streamlined, lighter and reliable over the years. In essence, they are a simple motor moving a pair of metal arms in a circle. Each arm carries a dead bird, which flaps as it circles and resembles a pair of birds landing, a pattern of decoys provides the backdrop and the feeding pigeon home in on the movement, swooping in to the pattern to provide the shooter with a variety of shots from all angles.

The 'Pigeon Magnet' over a stubble field: modern application of ancient decoying ingenuity.

There is a wide range of 'magnet' available at prices ranging from the very low ($120) to over $400. New varieties are constantly under development and include devices for four birds, those with two motors and two pairs of birds rotating in different directions and other devices such as peckers, flappers and cradles to show maximum variety of movement in the pattern.

For most purposes a standard, good quality device for two birds will serve well if used with a dozen or so shell decoys. Flock coated types are the best, as they do not shine in sun or rain. The Shoot Warehouse, run by professional pigeon guides and pioneering machinery inventors Philip and Will Beasley has everything you could possibly want made to high standards of quality.

A set of poles is essential for hide building and the best are double spiked for stability and telescopic for ease of transport and adjustability. You need six to make a good hide. Add to this list a good camouflage net, or nets of military type or similar and you have all you need to be ready to tackle the pigeons – except field craft and that you must learn; for all the devices in the world will not work without it.

Hammer guns for practical shooting

When I was a child of perhaps eight or ten years old, my father, a country GP, would occasionally take me on his rounds to visit patients in the rural farm communities of south Shropshire. Looking back I have a distinct recollection that in most of those farmhouses there was an old hammer shotgun propped up behind the kitchen door, dusty and neglected.

Later, I used to ride around with Ray Preece, the gamekeeper, in his old Simca van, which invariably had a rather nice hammer gun, with intricately patterned Damascus barrels, wedged between the front seats. He took it around while feeding the birds but left his 'better' gun at home for use on more formal occasions.

These memories stay with me because they illustrate how times and attitudes in some shooting circles are changing. A few years ago hammer guns were considered obsolete and many were neglected or misused to destruction, having been replaced by the more modern hammerless designs. However, a revival in the hammer gun has become well established and many sportsmen have rediscovered the delights of the external hammer.

Whilst hammer guns of many designs appeal to the collector and are discussed elsewhere, I will concentrate here on the hammer gun for use in a modern game-shooting context. We will therefore discount the non-rebounding lock, the pinfire and the percussion muzzle-loader. Instead we will consider the merits of the breech loading hammer gun with rebounding side locks in either bar action or back action form.

Hammer guns still perform in the field. Here is my 1885 16-bore Holland & Holland, having accounted for 35 pigeons during an afternoon in the pigeon hide – 119 years after it left the shop.

Guns of this type were made in various grades, cheap 'keepers guns' and 'export guns', through the well made but unembellished guns intended for the colonial market, right up to 'best' grade guns for the nobility and rich merchant classes. Game shooting in 1880 was little different from game shooting now – in fact the quantity of birds shot on most driven days now is much lower than it was then. As a tool for the job, the hammer gun is as suited to driven game now as it ever was.

Many people are not aware that hammer guns were made well into the 20th century. Many of the greatest shots of the Edwardian era preferred hammer guns to the new hammerless actions. Walsingham, Ripon and King George V all continued to have hammer guns made by Purdey after the perfection of the hammerless action and it is possible to find hammer 'Live Pigeon' guns made by the best makers well into the 1920s.

The Americans were way ahead of the British in reviving hammer gun appreciation. Two American authors quoted earlier in this book, Adams & Braden, famously espoused the qualities of best English hammer guns and their enthusiasm has proved infectious. Now the auction houses are full of American buyers on the hunt for hammer guns to take back to the States and sell for a profit at the various gun fairs and events held there. A quick search of gun dealers' web sites shows that dealers and shops in the USA probably have better stocks of British hammer guns that those in the UK.

I was once outbid for a very plain Purdey hammer gun at auction: by Cyril Adams. I thought he paid over the odds but I'm sure he got a good return on the other side of the Atlantic. Incidentally, Mr Adams deserves a mention as another of those excellent characters one rubs shoulders with at auctions, who is always ready to share his expertise and exchange a few words on a matter of common interest.

The exodus of quality English hammer guns to the USA is reaching epidemic proportions and it would be a pity to think that in a few years all of our quality hammer gun heritage will reside in the States.

Adams has pointed out that not many people realize just how few top quality hammer guns were ever made. The 'hammer gun era' lasted only about 25 years. Murcott's hammerless 'Mousetrap' was patented in 1871 and the Anson & Deeley hammerless boxlock (still made today) was available from 1875. Hammer guns went into decline in the last decade of the 19th century, once hammerless guns had been perfected – notably the Purdey and Holland

King George V using a trio of Purdey hammer guns. The King favoured hammer guns well into the 1930s, despite the general preference for hammerless sidelocks.

and Holland sidelock models that we know today. The combination of hammerless sidelock and reliable ejector really spelled the beginning of the end, though hammer guns continued to be made in some quantity until WW1. Consider how many hammer guns at this time were 'best' grade and allow for wear and tear, wars and neglect and what guns are left are relatively few in number, as Adams and Braden point out.

Purdey launched a new hammer gun in 2004, a snip at over $104,600 but again available just as it was in 1879 (with the modern addition of a safety catch), before Beesley invented the Purdey self-opening hammerless sidelock, which heralded its demise. AYA and Patrick Keen have both launched hammer guns in 2005, in response to resurgence in demand. This is the first serious re-appearance of the hammer gun as a mainstream option since before the First World War.

Genesis.....	HAMMER BREECH-LOADER ERA			Decline......
1861	1867	1875	1880	1884
Centre-fire Hammer Guns	Stanton Rebounding Lock Perfected	Anson & Deeley Boxlock	Purdey/Beesley Self-opening sidelock	Holland & Holland 'Royal'

This Pape hammer gun is of exquisite quality – proportion, engraving and finishing are all stunning.

Hammer guns for driven game shooting

The hammer gun defers to no other style of gun in balance, handling and elegance. It will shoot as far, as straight and as well as any modern gun and is a delight to use. Where it is at a disadvantage is in speed of loading. For driven shooting one has the option of the rare hammer ejector. These are hard to find and expensive but wonderful to use.

Some even have a self-cocking mechanism, where the fall of the barrels or (more generally) movement of the opening lever cocks the hammers. Each type should be carefully considered on its merits as some were rather experimental in nature and may not be reliable or safe today.

With practice, the normal non-ejector, manually cocked hammer gun is manageable in a hot peg for driven shooting. You need to be deliberate, purposeful, unhurried and rhythmic when firing, emptying and loading but your rate of fire will not be too much diminished, as long as you do not fumble or hurry your shots. As with everything, practice makes perfect.

Pigeon shooting & wildfowling with a hammer gun

I personally find the ejector an annoying extravagance when in a pigeon hide. The shells fly into the hedge and I have to dive into brambles to collect them at the end of the day. Not a problem with most hammer guns, as they do not eject.

The 'Live Pigeon' style hammer gun is excellent for shooting pigeon over decoys. It tends to be a little heavier than a game gun, often has a half pistol grip and filed rib and is of robust construction. It will absorb the recoil of heavier loads and take less of a toll over a day of shooting.

A bar action hammer gun of 7lbs with Eley Maximum cartridges will prove excellent for dispatching pigeons at good ranges and with the same load in Bismuth deals with ducks and geese in good style. I prefer a gun with a shorter stock than I use for game shooting.

This is because of the need to stand still as the pigeons come in to the decoys. I tend to use the Churchill style of shooting in the hide, shifting the weight to the side the bird breaks. The mount works better with a shorter stock.

Rough shooting

Hammer guns balance beautifully, are visibly safe when un-cocked and prove very pleasant when carried around. I use my Holland & Holland 16-bore hammer gun on many of my 'Walk and Stand' shoot days. It weighs a little under six pounds and shoots beautifully with 15/16oz of shot, perfectly adequate for fast partridges and pheasants out to 35 yards.

Maintenance

Hammer guns are relatively simple tools. Their longevity is testament to the excellent quality of materials and workmanship employed in their manufacture. After a wet day, stripping the mechanism, cleaning & drying is straightforward and satisfying. Anything that breaks can be filed up by a good gunsmith. The quality hammer gun will never wear out if cared for.

Caution – old guns!

Remember that most hammer guns are very old and they have lived long enough to be abused. Look for pitted barrels and check barrel measurements against the proof dimensions; they may have been lapped out to the point that they are on the edge of re-proofing becoming necessary.

Also be aware that hammer guns were made in some very poor qualities as cheap 'export' guns. Any that are still in circulation are likely to be in very rough condition by now and are best avoided. When buying hammer guns, as when buying any other old English guns, you should aim for the best quality you can find.

Not all English guns were made to the highest standard; there were many grades of gun produced for different markets. This is another reason why examining large numbers of hammer guns at auction is a good idea; you will become proficient at assessing the quality of individual guns and recognizing lesser workmanship when you see it.

Out-of-proof Damascus barrels may be salvaged now by the Teague 'internal sleeve' process, discussed earlier, which retains the external dimensions and appearance of the Damascus by lining the inside with new tubes and chamber liners.

As with any old gun, when buying a hammer gun,

Holland & Holland *Paradox*: not much of a duck gun

look for cracks in the stock or a poor wood-to-metal fit from a cheap re-stocking job. Also note that most people were smaller in the past than nowadays and you will come across a lot of hammer guns with very short stocks. Check the fit because long extensions can be ugly.

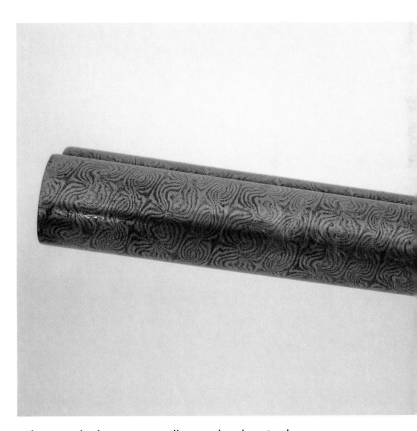

Those with sharp eyes will note the dent in these Damascus barrels, however you need a measuring tool to discover the wall thickness!

Useful hammer-gun points of reference

Rebounding Locks

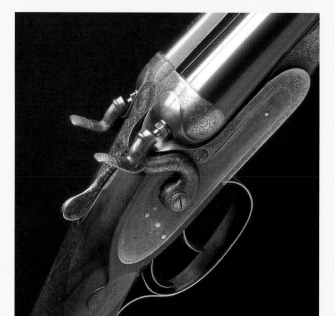

Non-rebounding Locks

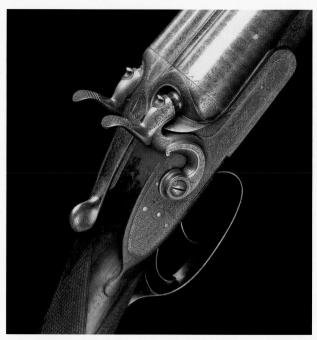

Rebounding locks are the preferable option when buying a hammer gun to use in company. The hammer will rebound back off the striker after the shot is fired, coming to rest in a position just short of contact. This allows the gun to be opened without touching the hammers after firing and makes it faster and safer to operate. (Rebounding locks were perfected by 1867 but not universally adopted immediately). The strikers are then withdrawn by small coil springs (cut-down biro springs are commonly employed for the purpose and work very well as replacements).

Non-rebounding locks have the disadvantage of the hammers resting on the strikers after firing. This means that the pins are sticking in the cap of the fired cartridges and they can jam when the gun is opened. To combat this, the hammers need to be pulled back to 'half cock' before opening. It is possible to inadvertently open the gun with a striker-pin sticking through the holes in the breech face, and the hammer bearing down hard on the opposite end. When a new, live, cartridge is inserted in the breech, if the gun is swiftly closed, the cartridge can (potentially) discharge unexpectedly.

Hammers

Hammers come in many shapes and sizes, some very beautifully filed and engraved. Generally, older guns have higher hammers resembling percussion era muzzle-loaders, later guns often have hammers that are below the line of sight when cocked.

BIG BORE WILDFOWLING GUNS

In times past, the early hunter of wildfowl quickly recognized the need to take shots at tough birds at long range. Col. Peter Hawker was using big-bore Manton guns before the 1840s to take coastal ducks and geese, when he was tired of shooting at puffins and larks.

Many 19th century sportsmen and market hunters used monstrous punt guns firing up to 20oz of lead to make multiple kills from large flocks but others, seeking portability, took long-barrelled shotguns of significant weight out onto the mud flats to ambush flighting geese and ducks, often in foul conditions. For this job, special guns were developed by the gunmakers of the time.

The obvious way to increase range is to increase the powder and shot load. While the 12-bore held sway at the game shoot, the salt marsh favoured the 2-bore, 4-bore, 8-bore or 10-bore. W.W Greener is often quoted for his advocated weight-of-gun to weight-of-shot ratio of 96-1. This means that to fire a 1oz load comfortably a gun needs to weigh 96 x 1oz (i.e. 6lbs). Big bore guns have to fire much heavier loads and, to absorb recoil adequately, they need to be correspondingly heavy. A single 4-bore can easily weigh 16lbs and fire 4oz of shot. Greener also advocated 40-1 as a formula for working out the optimal barrel length. Damascus barrels of 36-40 inches are therefore common in 8- and 4-bore. Some of these monsters are more like hand-held cannons than sporting guns and the complication of finding the right ammunition and carrying the great weight around helps to explain their demise in favour of the heavily-loaded 10-bore and 12-bore nitro magnums used by modern fowlers.

This Leeson 8-bore side-lock is of exceptional quality, built to fire 2oz of shot and sold for $17,800 at Gavin Gardiner Ltd in 2006 – a record for a gun of its type.

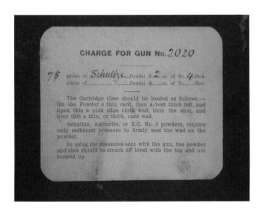

These case labels help to date the big Leeson and also provide loading data that will ensure the gun continues to be fed the correct loads.

Comparing the style and proportion – an 8-bore wildfowling gun (above) and a 20-bore muzzle-loading game gun.

Wildfowling guns developed distinctly from game guns as a branch of shotgun design in the last third of the 19th century. Game shooters were concerned with light-weight, rapidity of fire, elegance and ease of manipulation. Wildfowlers were concerned with chucking as much lead as possible as far as possible and expected long waits between each shot.

As game guns became modern top-lever operated bar-action sidelock ejectors, many large-bore guns retained earlier, less refined mechanisms; as the emphasis was on robust, reliable guns that could take abuse. The Jones under-lever with its rotary-grip of 1859, an inert but very strong system for locking breech loaders, was retained by many makers well into the 20th century, as it worked efficiently in dirty conditions and was almost indestructible. Hammer guns with back-action locks were commonly employed – because they enabled strength to be retained in the bar of

the action, without having to add weight to make up for the hollowing-out required to fit bar-locks or internal tumblers.

Most wildfowling guns were made plain. They were tools for the market hunter or the foreshore specialist and had to operate in very unforgiving conditions. Wood is generally unremarkable, with coarse checkering (to help muddy hands grip), locks are un-engraved or only marked with the simplest of borders and the odd scroll and all the parts are made heavy, for strength, and big, for ease of operation with cold, muddy or gloved fingers. Double guns are relatively common in 10- and 8-bore but single 8- and 4-bore guns are more so. Remember that many of these guns were working tools for market hunters, unlike the thoroughbred game guns of the wealthy.

However, big-bore guns also found favour with rich clients and often made interesting exhibition pieces. Highly decorated 4- and 8-bores will be encountered that were ordered by Indian princes in the 1880s and 1890s and others of exhibition quality occasionally turn up at auction and make very good money. Two come to mind that I saw sold at Bonham's in 2005: a beautifully-made Holland &

Holland single hammer 4-bore, that had come back from India, and a Westley Richards 4-bore, built on a boxlock action, that had been made for exhibition. If you want one of these you need to be thinking of $10,000-$16,000 a-piece. Plainer guns of good quality will make $6,000-$8,000 and very plain, rough old workhorses may be picked up for under $2,000.

The law

You can legally own large bore guns in Britain without a Shotgun Certificate as long as you do not own ammunition for them or try to use them. This has caused prices to rise in the last couple of years as they can now be displayed as ornaments rather than remain locked in a cabinet. This does not apply to 12-bore 3" magnums but it does apply to guns of 2-bore, 4-bore, 8-bore and 10-bore (depending on chamber length) and it makes no difference whether they are muzzle-loaders or breech-loaders or if they are hammer or hammerless. Americans are not permitted to use guns larger than 10-bore for wildfowling.

Ammunition

A lot of big-bore users hoard old cases and re-load them until they fall to bits. Those with the inclination to outlay some capital have brass cases made up to the correct dimensions and re-load them forever. Auctions are prime sources of used cases and vintage collections of un-fired cartridges and there is a healthy market for them every time they appear.

www.justcartridges.com stock 10-bore shells in non-toxic loads for wildfowling but anything bigger can be ordered from bespoke loader Alan Myers, who will make specific orders up according to the customer's requirement. One of the difficulties with large bore guns is the huge variety of chamber lengths and dimensions that are encountered. Some were made 'chamberless' for use with thin-walled brass cartridges, while others were chambered for thicker paper cases.

Now that lead is forbidden in Britain and the USA for wildfowling, fowlers need to consider the alternatives and bismuth or tungsten matrix are probably the kindest substitutes to use in old barrels and the best choice for the

Vintage cartridges like these Eley 8-bore gas-tight shells are always snapped up when they appear at auction.

The Vintagers of the USA

Since its founding in 1994 by Ray Poudrier, The Order of Edwardian Gunners, AKA 'The Vintagers,' has developed serious momentum in the United States. It now brings together people from all around the world to socialise and to compete with their vintage guns. The event has expanded from a single day to become a three-day festival for shooters and gun collectors.

As can be seen from the extensive list of competitions and rules outlined below, from the 2005 event, the whole thing is well organised, prestigious and attracts gunmakers, gun dealers, shooters and collectors from all over the world. Competition is fierce but as a magnet for all those interested in British double guns, it has proved to be an excellent meeting point for those of like mind.

Main Event

100 Sporting Clay targets at 16 stations

* HOA to have his/her name engraved on the permanent National Champion trophy and to receive National Side-by-Side Championship personal trophy.

* Trophies for top 10 places in Senior division, top 5 places in Veteran division, top 5 places in Ladies division, and top place in Junior division.

* A single trophy to High Hammer Gun.

* National/Corporate/Club award to Team with highest three-shooter aggregate for a team of up to four shooters.

* International Sporting Clays rules apply – low gun. An exception for ISC rules: 'spreader loads' of appropriate weight will be allowed.

* No changing of choke tubes allowed once the course has been started. Practice available starting at 8am.

Side Events

Shooting Sportsman Small Gauge Events 50 Sporting Clays Targets **16 ga** (1oz), **20 ga** (⅞oz), **24 ga** (¹³/₁₆oz), **28 ga** (¾oz), **32 ga** (⁹/₁₆oz), and **.410** (½oz) No tube sets or chambermates allowed for small gauge events. High score to win that gauge's 'Championship'. The next highest score of a hammer gun and a hammerless gun to win in that gauge's 'Hammer Championship' and 'Hammerless Championship'. Single high gun trophy 24 and 32 gauges.

Compak Championship 50 Sporting Clays Targets

Preliminary Event 50 Sporting Clays Targets

12 Gauge Hammer Event 50 Sporting Clays Targets

10 Gauge Event (not more than 1½oz) 50 Sporting Clays Targets

American Classics Event 50 Sporting Clays Targets Friday & Saturday, 8:30am to 5:00pm This event open to guns of American design only.

Paired Gun Championship Friday & Saturday, 8:30am to 5:00pm Shooter must have pair of guns and a loader. 200 targets in 4 min.

Two Man/Woman Team Championship Friday & Saturday, 8:30am to 5:00pm Two shooters to use 1 or 2 guns each, with or without a loader, to shoot 200 targets in 4 minutes.

Black Powder Compak Championship – 30 Targets Friday & Saturday, 8:30am to 5:00pm Cartridges must be loaded with Black Powder and not more than 1⅛oz shot.

Skeet and Down-the-Line Championships

Skeet competition will include 12, 16, 20, 28 and .410. 12 Gauge will shoot 100 targets, all other gauges will shoot 50 targets. Low gun (beneath arm pit). Delayed release up to 3

seconds in the pull. Option on first miss if straight option on Low 8. Squad to shoot high 8 then squad to shoot low 8. There will be a four-gauge Championship – 100 targets for 12 ga, 50 targets for 20, 28, and .410.

Down-the-Line: The main competition consists of 300 targets: 100 targets shot from the 16 yard line (singles), 100 targets doubles, and 100 targets handicap – men at 22 yards, ladies and juniors at 19 yards. Three points will be awarded for 1st barrel kill, 2 points for the 2nd barrel kill. Maximum score for the World Championship is 900 points. Note: Doubles will be scored as 3 or 0 for each barrel.

Separate 50 Target Events: Offered for 20 ga high gun champion, 28 ga high gun champion, and possibly a wobble trap champion (to be confirmed). **Tube sets are not allowed.** Maximum loads are as follows: **12 ga** - 1⅛oz. 7.5, **20 ga** - ⅞oz. 7.5, **28 ga** - ¾oz. 7.5 and **410 bore** - ½oz 7.5.

Trophies will be awarded for High Overall – World Champion (16 yd Handicap & doubles), Ladies High Gun – World Champion, Trophies for 12, 20, and 28 Champions and Certificates for Runners-up and High 16 yard, High Handicap, High Doubles, and Wobble Champion.

Small gauge 50 bird events for trap for 20 and 28 gauge events.

Ray Poudrier can be contacted at:

The Vintagers, P. O. Box 31, Hawley, MA 01339
email: vintagersray@hotmail.com

home-loader. Hevi-Shot is underdeveloped as a projectile at present. It is three times harder than your barrels and cannot be produced with uniformity of shape. If just one of these mis-shapen pellets finds a way through or around the shot-cup, it will score your barrel. At the time of writing, rumors are surfacing about a new version of Hevi-Shot emerging, that is softer than lead. This will have massive potential as a shot load for all applications and the only factor to then consider further is the price.

Alan Myers has thirty years experience in catering to the user of big-bore guns. He will make any cartridge to order, tailored especially for a particular gun so that chamber and bore dimensions match the ammunition exactly. For example, 8-bore loads can vary from 2oz to 3oz depending on the age and construction of the gun. He can provide nitro or black powder loads and plastic or brass cases. He will load a box of 25 or supply 6,000 in any bore size from .410 to 2-bore. A bespoke loading option like this ensures that big bore guns have a bright future and they have a devoted band of followers keeping them in service.

Shooting vintage guns: technique and style by Mike Yardley

'How does shooting a vintage side-by-side differ from shooting a modern Beretta or Browning?'

That was the question Dig put to me. I like to think that grace and elegance are important – what was once called 'good form.' But, what is it? Good shooting with any shotgun boils down to a few simple things – keeping the eyes on the bird, being in balance at the moment the shot is taken, and keeping the gun moving well. As with golf, cricket or tennis, one must follow through.

In Ripon's famously curt advice: 'Don't check'. The head must be on the stock as the trigger is pulled. The weight must be predominately forward (if you favour a Stanbury style stance). One must also use the feet well, stepping into the line of the bird whenever the opportunity presents itself. It's about economy of movement and effort.

Side-by-sides of the vintage era tend to be quite light compared to modern competition guns. In particular, they tend to have relatively light barrels (one of the defining qualities of a fine English gun). The use of the front hand to control the muzzles is critical. It becomes even more important with small bores that tend to start and to stop even more quickly.

I also believe that cartridge selection is an important issue. I favour light loads for side-by-sides. Excess recoil can be a real problem with a 6½ pound gun. If I am shooting clays, a 24-gram shell is ideal (for game with a normal 12, I favour a traditional $1\frac{1}{16}$ load). Now, I am sure Dig will present his own ideas, but I have had considerable success with the Express Lyalvale 24-gram load in vintage side-by-sides. In both plastic wadded and fibre form it is a cracking little shell and very smooth shooting. As I write this, Dig and I have just stepped off a range where we were (honestly) shooting 65-yard crossers with my old Lang hammer gun.

Another issue of particular importance to side-by-side users is upper body rotation. Light guns are not especially very forgiving to use. They accelerate quickly, but they stop much more easily than a gun of greater mass. The front hand provides fine control, but the power house of the swing – the engine of it – is the core of the body. Too many try and shoot with arms alone, or with poor body placement. This is a great mistake and typically leads to rushing and stopping – a poke and slash style of shooting. Dropping of the shoulder – which leads to missed behind and under – is a particular problem for people who have not learnt to use the body and feet well.

One often hears the advice not to cant the barrels. This can be confusing. What is critical with a side-by-side is that the barrels remain parallel to the line of the bird (an over-and-under should be perpendicular to it) – imagine a line drawn through the muzzle centres. To keep this parallel with the line of flight may *require* you to cant or twist the gun slightly as the shot is taken. But, the gun must not be canted relative to the line of flight as the shot is made. If it is, you will come off line and the wrong eye may take over if you are shooting with both eyes open. One tip for the right-handed, right master-eyes shot, when shooting driven birds, is to twist the gun anti-clockwise into the cheek as the shot just right of centre is taken. Otherwise this is a bird that can cause no end of problems.

Users of side-by-sides and light over-and-unders may also find themselves continually shooting in front of low driven game birds when they are 15-20 yards from the gun (and don't fool yourself: many, if not most, birds are shot at such ranges). The classic advice is that 'grouse wear spats' i.e. that it is easy to miss them above (which it is) but closer range partridge and pheasant are missed in front far more often than commonly realized. It is a major problem in my experience. Once the range is extended to 25 yards or so, the typical miss is behind (and usually due

Learn the 'line discipline' so essential in ensuring everyone's enjoyment of the day. This drawing from Lancaster's *The Art of Shooting* shows that issues have not changed over the years.

to stopping). High birds are missed behind because the range is misjudged. They may also be missed low and off-line as well.

Front hand position is another interesting issue. With an over-and-under, the hand should be in a middle position on the forend, but in the case of a side-by-side, the best place is at the front of the forend. It is a great mistake in my opinion to emulate the style of George V with an exaggerated front hand hold on the barrels. This is guaranteed to check the swing on many birds. An exaggerated front hold may, however, be used occasionally when one is using a gun which is a bit too short.

When using a classic side-by-side, it can also be a mistake to use excessive head pressure. Many, if not most English and Scottish side-by-sides have significantly tapered combs. Excessive head pressure may cause one to lose the bead – and, baffling until you realize it, it may also cause the wrong eye to take over because the eye that should be looking down the rib, ends up staring into the breech.

I do not favour a very deliberate technique with side-by-sides inside 30 yards or so. Two-eyed shooters should put all their energies into sustaining focus on the bird and keeping the gun moving. Once one has trained, lead should be applied subconsciously as advised by Mr Churchill. Beyond 30 yards, or in difficult conditions (for example, when the wind gets behind a bird) a more deliberate approach may be required. I have no problem in telling someone that they should be three/four yards ahead of a long bird. To those who say one man's inch is another's yard, I would reply: 'only if he is looking at the barrels.' Some people may, however, relate better to gates, trees or cars than specific distances.

Should one swing through or maintain a lead? Both techniques can be useful. Swinging through works. Maintaining a lead can be instinctive in some situations but can lead to misses in front or off line. I favour a simple approach where one takes the barrels to the tail of the bird initially and then pushes forward, rotating the upper body and maintaining fine control and elevation with the front hand. I believe it is important to create a muzzle target relationship as quickly as possible but I do not favour any deliberate tracking (except when teaching beginners).

I have searched for the right words for many years – 'point and push' sums up my philosophy as well as any. And, most important, ***don't be in a hurry***. Good shooting is unrushed and rhythmic. If you are musically inclined, one might note that all shooting should be conducted to three beats –   ONE: TWO: THREE….

BUM: BELLY: BEAK….

YOU: ARE: DEAD.

But the tempo changes depending on the speed, angle and range of the bird. Keep your eyes on the bird, keep your head on the stock and keep the muzzles moving.

Shooting with vintage guns – etiquette and safety

A vintage breech-loader is equally suited to the shooting of driven game as a modern shotgun. After all, it was designed for that very purpose. The 2nd Marquis of Ripon was still successfully shooting his Purdey hammer guns with non-rebounding locks in 1909, (when he had them re-barrelled and choked extra-full and extra-full) and continued to do so until 1923, when he dropped dead in the heather. However, changes in attitude over the last couple of decades have led to far more over-and-under guns being used for formal game shooting and it is fair to say that the over-and-under is now a more common sight at a game shoot than the side-by-side.

It is possible that shooting companions not familiar with the sight of vintage guns, especially hammer guns, may have concerns about the safety of these old weapons. It is incumbent upon the user of the vintage gun in polite company to make sure his companions are put at their ease with regard to safety issues.

If shooting a hammer gun, make sure that the gun is open when in company, as one would with a hammerless gun. Although a hammer gun cannot be fired, even if loaded, unless the hammers are at 'full cock', your companions may not know this and will be disturbed by your moving among them with a closed gun, even if it is visibly safe to you.

Owners of hammer guns with non-rebounding locks are advised not to use these for driven shooting unless very experienced handlers of such guns. The excitement of a hot peg can lead to hurried reloading; and the need to withdraw hammers to half cock before opening the gun

can easily be forgotten in the rush. This is not safe and accidental discharge, though unlikely, is possible and a tragedy could occur, or at the very least, an embarrassing incident that will make you no friends. You could also damage the strikers.

The alternative is to shoot with an experienced loader, who will have more time and less adrenaline to contend with. However, a word of warning about using loaders: many shoots will have a policy on loaders or 'stuffers' and will not allow them. Although the rule may have been made assuming that all Guns are using modern ejectors, the fact that you are using a non-ejector hammer gun and have a slower rate of fire will not always be accepted or understood.

I recently had to load for myself and decline the services of my companion, who was planning to 'stuff' my hammer non-ejector for me, because a surly member of the shooting party commented that it was my choice to handicap myself and the rule was 'no stuffers'. Now the shoot captain would have taken my side, of this I have no doubt, but I have no wish to be confrontational on a day that is supposed to be sociable.

We must remember that we single ourselves out for attention when we choose an old gun to bring to a formal shoot. Just as the man in flamboyant and unconventional attire must shoot like God in order to avoid ridicule, so the user of a vintage gun must be exemplary in all manner of safety and etiquette, as he will be drawing unusual levels of attention to himself.

If planning to shoot with black powder, always ask the shoot captain if this is permissible in advance. Few will object but all will appreciate the thoughtfulness involved in making the request. It is always better to ask – he will then be likely to explain to the other guns what is going on before the start of the shoot. People deal with the expected better than the unexpected.

Always be courteous and offer to desist from using the black powder should any fellow guns find it objectionable. This way you will be treated with kindness and tolerance by most. If your fellow Guns think you are imposing your eccentricity in a way that makes them uncomfortable, they will resent you and a lot of unpleasant muttering may be the result.

Modern guns tend to be proved for the 70mm (2¾") cartridge with a service pressure of 3¼ tons per square inch. Many vintage guns will have been proved for the lighter 65mm (2½") cartridge with service pressures of only 3 tons per square inch. Make sure you have the right ammunition and plenty of it. Your companions may be willing to lend you (or sell you) cartridges but they may not be safe for use in your gun. If you are using more than one gun (whether of different bore size or different chamber length) keep ammunition separate – in separate and distinct bags if possible. Never mix cartridges of different sizes in pockets or bags – you only need to get it wrong once!

Make sure your guns are serviceable before you use them. Check that the gun is 'in proof', that the ammunition you are using is suitable for the gun and that the mechanism is sound. It is good practice to have a competent gunsmith check your gun before using it for the first time. Even if you are competent yourself, a second opinion is a good 'safety catch'. We are all human and we are all capable of overlooking the obvious. Have someone measure the bores and the barrel thickness, check there are no dents or dangerous pits or bulges and that the barrels are 'on the face' of the action.

It can be a good idea to show your gun to those in line next to you. You can do this in the gun bus on the way to the first drive. Show them it is in good condition and that you know what you are talking about. This will reassure them that they are not shooting with some lunatic whose barrels are likely to come flying in their direction (in pieces) half-way through the drive. Remember the old adage; *'Ignorance, prejudice and fear go hand in hand'*: if your companions understand you and your motivation and accept your competence, they will be comfortable shooting with you and your funny old guns. They will often be interested and may even catch the bug! On the right are illustrations of safe handling of hammer guns with rebounding locks.

Safety and Vintage Guns

Recent discussions and correspondence with vintage gun enthusiasts in the USA and the UK show a determined following for old guns in the field in both countries, though both have very different shooting traditions and the conditions in which shooters may find themselves using old guns will vary a great deal.

This makes it difficult to generalize about what is safe practice and what is not. However, there are certain issues relating to the safe use of different types of vintage gun in the field that are usefully brought to the attention

Right: Peter Jackson demonstrates safe-handling of a hammer gun with re-bounding locks

1. Insert the cartridges

2. Close the gun, hammers not cocked

3. Raise muzzles skywards before cocking the hammers

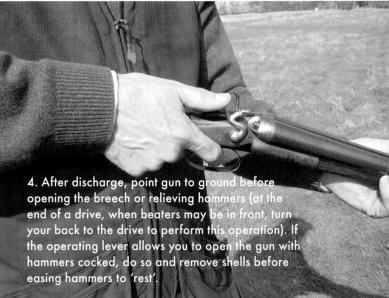

4. After discharge, point gun to ground before opening the breech or relieving hammers (at the end of a drive, when beaters may be in front, turn your back to the drive to perform this operation). If the operating lever allows you to open the gun with hammers cocked, do so and remove shells before easing hammers to 'rest'.

1. With hammers at 'half cock', open the gun and insert shells.

2. Close the gun, muzzles pointing downwards, hammers at 'half cock'.

3. With barrels raised, pull hammers to 'full cock'. The gun is ready to fire.

4. After the drive, release the hammers to rest and pull back to 'half cock'. The gun may now be opened and live shells removed. If the operating lever allows you to open the gun at 'full cock', do so and remove shells before easing the hammers.

of the user. I will explore some of these but add the caveat that there is no substitute for common sense and muzzle control. Whatever the action, if your muzzles are always pointing in a safe direction, you will endanger nobody.

Safety Catches

Hammerless guns suffer from a major defect: they are always cocked when they are closed and a fall of the tumbler will cause them to discharge. For this reason, gunmakers began to include safety devices in their hammerless guns.

In many cases, these safety catches are no more than trigger locking devices. They stop the trigger being pulled and therefore prevent discharge of the weapon by conventional means. They do not stop the tumblers from falling should the sear slip from the bent. A worn sear, an oil-clogged bent, a jolt or metal fatigue may all occur and cause the gun to discharge even when the safety is 'on'.

Some safety devices fitted to better quality sidelocks and boxlocks include what is generally known as an 'intercepting safety'. This usually involves a second sear, or a bolt, which prevents the tumbler falling unless the trigger has been pulled. These are undoubtedly safer than simple trigger-locking safety catches but Greener warned in *Modern Shotguns*, in 1888, that there have been circumstances in which even intercepting safety devices have failed.

Hammer Guns

Hammer guns do not generally have a fitted safety, though one was fitted to the 2004 Purdey hammer guns that were made in a limited number in that year. The 'at rest' position of the hammers actually provides a greater safety option than a hammerless gun, which is always cocked when closed.

Hammer guns with rebounding locks in good condition are incapable of firing unless fully cocked. When closed and in the 'at rest' position, the gun will not fire. Hammer guns are therefore visibly safe when the hammers are at rest and may be carried safely closed and un-cocked. Where you consider it safe to carry a hammerless gun closed (and therefore cocked) with the safety on, it is actually safer to carry a hammerless gun closed and un-cocked.

Hammer guns with non-rebounding locks will not theoretically discharge a cartridge from 'half cock'. The hammer must be withdrawn to 'full cock' before it can

Left: Mike Yardley demonstrates safe-handling of a hammer gun with non-rebounding locks.

be fired. There is some dispute as to the veracity of this in some quarters.

Greener offered the following advice on the carriage of hammer guns in *The Gun and its Development*: '*Hammers should never be left resting on a cap or striker when the gun is loaded; let the hammers be carried at full cock*'. So, it would seem our forefathers were at ease with the practice of carrying a hammer gun ready to fire. Many today would disagree. Certainly, guns with non-rebounding locks should be carried at half cock; never with the hammer in contact with the pins.

When shooting driven birds from a peg, hammers may be fully cocked immediately, with the gun held at 'ready': muzzles skywards and facing the drive. When walking with the gun, for example walking up stubble fields in England, one must exercise judgement about whether to have a rebounding hammer gun closed and cocked or at rest.

Walking steadily on even terrain such as open stubble fields, away from others and anticipating a shot at any moment, I personally believe that a hammer gun may

reasonably be carried closed and cocked. Some shooters prefer to carry the gun with hammers at rest (or half-cock in the case of non-rebounders) and sweep the hammers back when the gun is mounted. I have never found this works and worry that an incomplete 'sweep' may lead to the hammers falling prematurely. When walking on rough terrain, I carry the gun closed and un-cocked and limit myself to cocking only the right lock when a bird flushes and taking one shot. If a flush can be anticipated, when walking up over dogs for example, one can often stop and cock both locks immediately preceding the flush.

It has been suggested by some that a hammer gun is best carried with both hammers cocked and the breech open, cartridges inserted. This way, upon flushing game, the gun may be closed and raised, ready to fire with ease. This raises a few objections; on guns with inert closing actions, I fear most would shoot very little game if using

The author's 1885 Holland & Holland 16-bore still works as well today as when it left the factory.

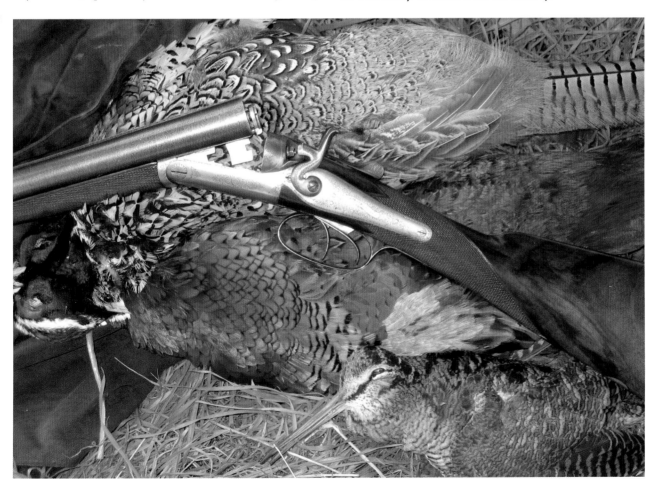

1883 Holland & Holland rook rifle converted to .410 at a 'Sporting Clays' shoot. These also make excellent rough shooting guns and can be used by youngsters. They offer high quality and no felt recoil.

this method, though with snap-actions it may work. There is a possibility that raising the muzzles in a hurry to take a shot may jolt the sears out of the bent and cause an unexpected discharge, though this is unlikely in a quality gun in good condition. I am also wary of carrying a gun muzzles-down in rough terrain, as they are at risk from dirt or leaves entering from the ground un-noticed.

A friend pointed out to me recently that in a world where people have common sense, rules are unnecessary but in a world where there are rules for everything but no common sense, we are in constant trouble. Exercise common sense whenever you use your old hammer guns.

Vermin Shooting

Vermin shooting gives you a great excuse to bring out the odd guns that you do not feel quite right for driven days or pigeon shooting, where efficiency and speed of shooting require guns with modern performance, or something approaching it. It is a chance to have some fun with the more quirky contents of your gun safe.

I have a cheaply bought Adams & Co back action hammer gun. This was a decent quality gun in its time but has seen better days. It has been sleeved – not well sleeved as it happens. The barrels when I got the gun were 25" long, heavy and heavily choked (full & full) with 2¾" chambers. In fact the sleeving job had not been properly finished, as the muzzles were still sharp and cut off not quite straight!

Luckily the joint between old and new tubes is good and tight and the blueing is tolerably well done. The gun actually shoots well; I have used it in a pigeon hide to good effect several times and have even shot driven partridges with it and once used it for a round of Skeet. However, I have since modified it. It has now been opened out to Improved Cylinder in both barrels and will be used with subsonic training cartridges for close range bolting rabbits. It is rough enough not to worry about knocking it on a ferret box or laying it on the ground, strong enough to take the rough use it will get and it still has lots of style.

For the occasional squirrel shoot we organise, I like to take out the odd single barrel hammer gun for an airing. They are slow to load and not the thing for general use but I still enjoy hearing them do their thing, 120 years after they first went into service. They also make a good companion for walking the hedgerows in search of summer rabbits or the occasional snap shot at a pigeon clattering out of a hedge.

Like many, I began shooting on a budget of nothing and my gun collection started as a motley group of things my friends found lying around their farm outbuildings. I remember an old 'Gem' air rifle that someone found under a water butt. That needed soaking in engine oil and freeing up. I eventually passed it on to someone more interested in air guns than I.

Another was an old 9mm Webley and Scott bolt action. I took it to bits, stripped the stock and re-finished it. I was probably 13 at the time but I remember the satisfaction of making it work like new and bringing the figure of the walnut stock out to see the light of day again, after decades hidden under dirty varnish.

This little Webley was my ratting gun. Better and faster than an air rifle in a dark cowshed – on go the lights, swing onto the beams and knock down a fat rat before they all disappear into the roof space.

We all talk about memorable shots; that high curling January pheasant or a spiralling teal on the wind. Well equally fixed in my memory is the rat that ran out of a grain sack like a bolt of lightning, over my foot and down the stairs of the outbuilding towards the yard. I spun and fired in a moment of pure instinct and ratty lay dead on the bottom step, shot through the head with my tiny charge of No.6.

I even used the 9mm for rabbit shooting in the summer. Evenings creeping along hedgerows to ambush the young bunnies were often productive, even with an effective range of little more than 10 yards. These simple pleasures with inefficient old guns still appeal to me. It is not just the 'best' offerings of the top London makers and days after pheasants that make up the Practical Eccentric's shooting diary, worthless old nails and lowly quarry have a charm all of their own.

Records

A gun collection will be more rewarding if you keep good records. This way you can keep an eye on what each gun has cost you in repairs and maintenance, the details are there for reference in case of loss or theft and it will be interesting in the future to see what you have owned and done over the years. As you discover more about the gun or the maker, you can add this to the record and it will be easy to use for reference; all the information being in one place.

A computer database is a good start, is easy to access and update and requires no storage space. The following format (in the box) is one I have found works well. All you need is a basic laptop PC and a printer/scanner. These machines are now available for under $200 and enable you to put photo images into your database and print them at will.

Security

Security is a fact of life for the gun owner. At the very least you need a steel safe to lock away your guns, window locks and door locks, trigger locks for travelling and a secure car

A typical steel security cabinet. They can be disguised in wooden cases to make them more attractive and less obtrusive. British law is very strict of the security of firearms and police approval must be had before a certificate will be granted.

Shotgun Register

Maker Holland & Holland

Serial Number 8749

Name on Locks Holland & Holland

Name on Barrels Holland & Holland 98 New Bond Street London

 Winner of all the Field rifle trials 1883

Gauge 16

Action Type Back action sidelock

Engraving scrolls and border

Date of Manufacture 1885

Barrel Material Damascus

Barrel Length 30"

Choke I.C & ¾

Rib concave game rib

Stock Type straight-hand French walnut, ebonite extension

Stock Length 15"

Fore Part Type splinter, Deeley & Edge catch

Proof Marks 3 tons, 2½", BNP, 677, NOT FOR BALL, 16

Purchased from Mike Yardley

Date of Purchase September 2003

Purchase Price $3,440

Sold to

Selling Date

Selling Price

Work carried out during my ownership

- New hammer pins – August 2003
- Re-finished the stock after a very rainy driven pheasant day removed the French polish – November 2003
- Put striker springs in (adapted from a biro) – August 2004

Notes Stock has little cast. Eley Grand Prix cartridges seem a little fierce. Express gentler. ⅞oz sufficient for all shooting. Used on pigeon, driven partridges & pheasant. Not a good gun for the hide or sporting clays. Needs to be swung freely. Excellent for walking up game. Mike Yardley used it to win the 16-bore hammer gun class in the British Side-by-Side Championship 2004

A few pits in both barrels, not serious.

November 2004 – got it cracked! 7 wind-blown fast partridges at every angle for 8 shots. Leave the stock alone.

boot for when you have to leave the gun during a journey. If leaving your gun in a car boot, remove the forend and take it with you or lock it in the glove compartment. This will make it useless and virtually impossible to sell and thus less attractive to a thief, should the car boot be broken into.

The Gun Room

The country house had a library, a dining room, a drawing room and it had a gunroom. Few of us these days have the space or income to claim all of the above and our guns are invariably stored in parts of the house that serve dual functions. Mine doubles as my study. The room is small and crowded but it is personal and contains my curiosities, my books, guns and cleaning equipment, my ammunition and shooting clothes. In short, all the things my girlfriend would rather not have on display in the rest of our home.

Actually, this is a blessing rather than a curse and my gunroom has become my retreat. I am writing this book in my gunroom, surrounded by my things, at my father's

The Gun Room Library – my top ten

1. *The Modern Shotgun* – three volumes (Major Sir Gerald Burrard)
2. *The Gun & its Development* (W.W. Greener)
3. *Amateur Gunsmithing* (Mills & Barnes)
4. *In the Gun Room* (Major Sir Gerald Burrard)
5. *The Shooter's Handbook* (Richard Arnold)
6. *The Shotgun – History & Development* (Geoffrey Boothroyd)
7. *The Shotgun* (MacDonald Hastings)
8. *Lock, Stock & Barrel* (Cyril Adams & Robert Braden)
9. *The Shotgun: A Shooting Instructor's Handbook* (Michael Yardley)
10. *British Gunmakers Vols 1 &2* (Nigel Brown)

old Victorian desk. On the shelves are the ephemera of a gun enthusiast – an old German sniper's telescopic sight, a broken sidelock's buttstock of fine figure, my books for reference, some ancient leather cartridge magazines and gun slips, cleaning rods and a modest trophy or two. On the walls are photos of family, college rugby XV's, variously collected edged weapons from my travels, old air rifles from childhood collections, trout flies and an opium pipe. These are my comforts; this is my space.

Time spent in the gun room is relaxing and contemplative. Cleaning guns, cleaning boots or adjusting or disassembling some curio provides us with time to be focussed and active but it allows our brains time to recover from the stresses of our working lives and our normal responsibilities.

The gunroom needs to contain books on guns to be consulted, read and re-read. I have asked some of the most engaging shooting men I know to provide their ten favorite books on shooting and shotguns. By doing so, I hope the nucleus of a shooting library will emerge and provide the reader with historically interesting and informative perspectives on our sport as well as practical technical reference sources to help him understand and enjoy his shotguns, their mechanics and their history. Here is our list of books for the gunroom:

Becoming a shot
How does a person become involved in shooting?

Consider this: most young people have no land, no money, little or no equipment, no contacts, no track record and few developed skills. How does such a person find a way into shooting?

Everywhere I read articles in the shooting press about the young and how important it is to get them involved in shooting sports and the countryside. There are discounted 'Young Shots' days at some of the major shooting schools and there are lots of 'Colts' classes in clay competitions. In fact, the organiser of one *Sporting Clays* shoot I often attend makes a point of giving a small 'runners up' trophy to every entrant in the under 16 category by way of encouragement. These details matter; small things make a difference when you are young.

Rufus Dexter (age 11) with gun and squirrel. Many urban teachers lack understanding and tolerance when faced with shooting and young people.

However, how many of us can honestly say we *actively* encourage young people to shoot with us? We may bring a keen young family member along, but the kids of shooters will often be the converted and the privileged and will naturally have access to shooting anyway. What of those youngsters outside the usual shooting circles? If we are to promote awareness of field sports, we need to promote it 'outside the fold' for it to have any effect. My interest in shooting and the countryside was developed in childhood, through access and experience. This is the key time.

We may not like to admit it but shooting people are often a conservative lot and though a warm and friendly crowd when together, rather jealously guard their sport. How many reading this would honestly respond positively to an enquiry from an unknown teenager asking permission to shoot vermin, use an air-rifle or shoot pigeons on their land?

As a teenager, I secured the kind permission of an old farmer to shoot his rabbits, only to be told after a few visits that I was no longer welcome because '*Mr Pilchard told me you would shoot my sheep*'. 'Mr Pilchard' (not his real name) was a local publican who also shot adjoining land and clearly did not welcome my presence. Hardly an encouraging story but I expect it is not unusual.

My experience is that most politely written enquiries about some free rough shooting will be met with either no reply or a negative response. This negativity has serious consequences for the public perception of shooting sports later on. Some of these kids will persevere and by hard graft, persuasion and effort, find what shooting they can and build up a number of useful venues to frequent. Others will be put off and may never shoot seriously, they will also retain a negative view of the shooting fraternity as a closed, inward-looking club. This is awful PR for shooting sports and in time every teenager 'lost' to shooting could be an adult willing to see it banned in future rather than one actively involved and fighting to promote understanding and tolerance.

Now, I am not suggesting that every gamekeeper should give *carte blanche* to every tearaway with an air gun wanting to run around the woods or that every farmer will be comfortable allowing all the local kids free rein on his land. There are many good reasons why one would not want to do either of these things. What I am suggesting is that everyone keeps an open mind and looks upon a request for some shooting as a positive thing and tries to find some way of harnessing this enthusiasm.

Perhaps offer to take the enquirer with you when you set and check your traps, offer advice, assess the person concerned and see what level of trust you can extend to them – and respond when that trust is earned. Offer some beating or working party involvement, with perhaps the sweetener of a day out ferreting or roost shooting under supervision.

It is really not that inconvenient and it is often very handy to have an enthusiastic companion to help. All this will serve to develop understanding. You will impart some of the knowledge you have learned over the years and the keen youngster will absorb it with amazing speed and come back wanting more.

When you can trust the person concerned, when you know they are responsible, aware of what is permissible, what is necessary and what is forbidden, allow a little more freedom. Perhaps offer the run of a few fields to shoot evening rabbits with an air rifle, an outing to assist with lamping foxes or a bit of decoying in a stubble field.

At worst you will have helped to develop a future

voter with a lifelong understanding of the nature of country life and some empathy for those involved. You may even have furnished the world with a little more awareness; a little less ignorance of who we are and what we do.

At best, you will have another trusted pair of eyes watching your land and reporting anything amiss: poachers, dumpers, trespassers, vandals, inconsiderate dog walkers, broken gates, holes in pens, fox activity or anything else. Young people have a lot of time on their hands and get very involved in their hobbies. They become very observant and aware of things in their world.

You will get free labor because they will love helping you out around the shoot and when they are old enough, earning money and settling, they will join a shoot and regenerate it with new blood as we, the older generation fade into history. If we all try to do this, our sport and its future are in safe hands. If we do not, the increasingly urbanized, sanitized version of life that most people live today will totally eclipse any alternative.

Ignorance of shooting and countryside management will lead to the uninformed urban majority banning everything they do not understand and we will be blown away on the wind, like the smoke from the last shot fired at driven game. That time will surely come if we do not actively educate and encourage the next generation. We must start now.

Consider what you can do. Don't leave it to the shooting organizations to do it on your behalf. Think of someone you know, not involved in shooting. Invite him to join the beating line with you or to sit with you in a hide and try to hit a pigeon or wait together at the water's edge for incoming duck. Introduce him to the dogs, the pickers-up and the fun involved in a shoot day.

Have him come along to the clay ground next time you go and have a go with a gun. Find a less than challenging stand and let him taste success and leave him wanting more. Shooting is an easy bug to catch; you just need exposure to the infected.

If you know a youngster keen on shooting but without the funds to buy a gun, offer him one you don't use. Most of us have at least one gun in the cabinet that we have not used for months, if not years. Is this not better employed in the field in the hands of an enthusiastic teenager? Teach him to clean the gun, to assemble and disassemble it and coach him in safety. He will never forget you and if the sport is in his blood, he will never look back.

I make a habit of buying certain types of gun at auction when I find them. They have to be cheap, simple, solid and characterful. Recent examples were a Cooey 12-bore single, a Webley & Scott .410 bolt action and a Harrington & Richardson .410 single. I paid $40 for each, stripped and cleaned them and gave them away. The recipients were people new to shooting with little money to invest but plenty of enthusiasm, or youngsters who were as pleased with their new gun as I would have been at their age.

A shooting friend with rather more money than me, a one time top-flight footballer, now lucratively engaged in the entertainment industry, came duck-flighting with us one evening. He confessed in the pub afterwards that he could not lend his pair of 20-bores to his father for the coming weekend because he had been shooting in Devon.

The story emerged that during the shoot, a teenage beater had been very taken with the guns in question and,

A burst barrel – cause unknown but probably due to a blockage. This accident occured in the firing hole of a London gunmaker.

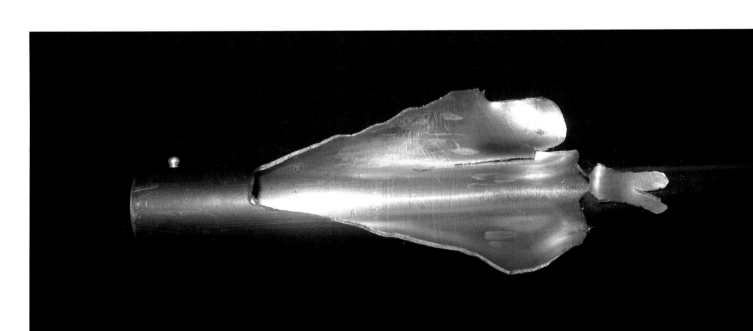

touched by the boy's enthusiasm, my friend had given him one of the guns and asked the boy's father to make a donation to Harefield Hospital, of what he could afford, in return. We cannot all afford such largesse but these gestures, at whatever level you feel able to contribute, go a long way to secure the foundations of the future of shooting. Do your bit.

Safety and young people

When I wanted an air rifle, aged eight, my grandfather got word of it and handed me a little booklet on shotgun safety: *The Gun Code*, published by the British Field Sports Society. The cover had *'Terence from GCWW'* hand written in blue ink. My great uncle Geoff had given this to my uncle Terence when he was in his teens and here it was passed on to me.

Inside was a brief note from the then Home Secretary, Henry Brooke and this simple addition from WRB Foster *'This book has been written in the hope that it will help all of us shooting men to become what in our hearts we would like to be, namely good sportsmen and safe companions'*. Not so much changes with the passing of years

My first gun-safety book – learned by heart as a child.

after all. We could say the same words to a prospective Shot today and they would be just as relevant.

The deal was simple: If you want a gun, you have to be able to quote every word of this booklet when quizzed. Then we'll think about it. Of course, I learned it by heart, in the way that single-minded eight-year-olds can when they want to, and so by the time I got my gun I was well aware of the importance of safety.

It was made clear to me that *one* transgression regarding safety that came to my father's attention would lead to the instant confiscation of my gun. No second chances, no warning, no mitigation, and no argument. I had to be *safer* than everybody else – all the time. I made sure that I was.

Of course, safety is *the* issue. Everything else depends on it, and without it nothing else matters. We have a duty when teaching the next generation to equip them for a happy life in the field; alone and in company. An unsafe Shot is unwelcome and unloved, a danger to himself and others. We must ensure our protégés become respectable and loveable – and safe.

The shooter on 'L' Plates

'The best evidence that a boy possesses the right material is an early craving after toy guns and swords.' Owen Jones *The Sport of Shooting* (1911)

Little boys like toy guns but must learn to distinguish between the war games they play and preparation for the shooting field. There exists today a popular brand of child-size replica shotgun. It can be found advertised in the classified section of the shooting press. It fires caps attached to mini cartridges and is designed for training kids to shoot. It resembles a normal side-by-side and comes with a gun slip and cartridge bag. If you have a keen youngster and are fortunate enough to shoot driven birds, take him along with the replica but insist that he treat the thing as real at all times.

Stand him at your peg and allow him to un-slip his gun, check the tubes are clear and load. You will wait together at the peg and he can mop up all the birds you miss.

Watch him carefully. Be firm and disciplined and serious in the treatment of the gun; exactly as you would be if it were real. Make him aware of the muzzles at all times and never tolerate slackness. No swinging across the line, no sloppy stance and no walking around with cartridges in the breech is to go uncorrected. Make him sleeve it between drives and stow it properly in the transport. Have him open the gun as it comes out of the slip and check it is clear. Let him mingle as if he were one of the Guns and expect him to observe the same discipline.

VINTAGE GUNS FOR YOUNGSTERS

This is a matter of choice. There are many good reasons to go for a modern gun when you choose one for your child or teenager. The cheap, reliable, easy maintenance over-and-under in 28-bore is probably the best choice in most cases. It will be easier to shoot and cheap to buy. It is an eminently sensible choice to make for a first gun. However, if a traditional gun is one you are considering, history provides us with some interesting options.

Our first gun is something we spend a lot of time with. We learn about what makes it work, how to treat it, what it is able to do and what is beyond its capabilities. We get to know its lines and balance and the grain of its wood and shape of its levers and joints. Give a boy a nice quality thing and he will learn to appreciate it and care for it. The impression will be lasting. Geoffrey Boothroyd once wrote of an ancient Jones under-lever operated hammer gun of forgotten make, which was his first gun. Considering Boothroyd's distinguished contribution to the cause of the vintage gun later in life, the impact was clearly profound. It really depends whether you want to approach the subject as equipping the boy for a sport or instilling in him a value of the beautiful things that will be his shooting companions.

Of course, you should also take into consideration the opinion of the boy concerned. If all his friends have competition guns and he wants to be like them, he will not be impressed with Dad's insistence on shooting with 'old junk'. Better that he gets a gun he wants to shoot with than is put off by having what you want him to have.

Choice of bore

Traditionally guns for boys were made in smaller bores. It is possible to find smaller sized guns, both hammer and hammerless, in .410-, 28-bore, 20-bore (and less commonly the obsolete 24-bore and 32-bore).

The great Percy Stanbury instructing a youngster (one with an eye-dominance issue).

I have been advised by those better informed in the art of shooting instruction than me that to start a boy with a small bore is a bad idea. Small bore guns are harder to shoot well and make shooting more difficult. I accept this argument entirely – for clay shooting.

For game and rough shooting, a boy must learn the *effective range* that his gun has in his hands and learn to use it accordingly. He must learn field craft to bring his quarry into shot and take only those shots he can be confident of hitting. With experience, he will get better. I now shoot rabbits with a .22 rim fire but my early days with a medium-powered air rifle did me no harm, indeed they taught me how to stalk my prey, read the body language of rabbits and pigeons in various states of alarm and learn to predict their flights and runs.

In my opinion the .410 offers no advantages over the 28-bore. If possible, go for the latter. The 28-bore is a very useful gun, capable of effective use in all game shooting

Single-barrel guns were commonly cheap utility farmer's tools. The picture shows a Hughes 16-bore c.1880, a 1960s Webley & Scott bolt-action .410, and a Harrington & Richardson .410 c.1900.

contexts in skilled hands. In the hands of a learner, it provides realistic performance in a light package. Interestingly, it is a little-known fact that the 28-bore produces the best, most even, patterns of any of the popular shotgun sizes. It also has a strong following among serious game Shots – the late Sir Joseph Nickerson performed famously with a trio of Purdey 28-bores in all game shooting situations. It is today an increasingly popular gauge for high-volume shooting such as doves in Argentina provide, where upwards of 600 shots may be fired in a day. Needless to say, it brings those doves down with regularity.

It is important to remember that small-bore guns will recoil heavily if loaded with heavy cartridges – this is off-putting for youngsters. Start with light loads and never exceed what the gun will comfortably handle. Greener used to say that a gun should weigh 96 times the maximum charge it fired (the 96-1 rule). Though some contemporary writers argue with this in the context of modern guns, it is a useful rule of thumb. The old gunmakers knew a thing or two and their advice is worth heeding.

Do not give a young shooter a gun that he cannot comfortably handle due to weight or length; it will only serve to dispirit him. However, as soon as he is able to handle it, give him a 16-bore or 12-bore, with light cartridges and get him used to the feel of a full-sized gun.

Single barrels

Despite what I wrote in the previous paragraphs about the superiority of the 28-bore, they are hard to find second-

hand and often expensive. If you have little to invest but want a lively gun of solid construction, get a Webley & Scott .410 bolt action. This is the advice Mike Yardley gives on the subject in his 2001 handbook for shooting instructors *The Shotgun* and it is hard to disagree. Such guns can be had for very little (I just got one for $60 with an air rifle thrown into the bargain) and they are almost indestructible yet well balanced; and they have character.

Side-by-sides

The children of the privileged classes were evidently indulged in the past, or so the evidence of guns built to dimensions suited to children would suggest. It is not uncommon to discover good quality side-by-side guns in smaller bores, with small, neat actions and proportionately short barrels and stocks.

These apparently come up for sale quite regularly. I expect to find at least one or two each time I go to an auction viewing. They are bought for a child of a certain age and when out-grown, sold to finance the next size up. This is one way to introduce a child to shooting. The guns are delightfully made as scaled down, balanced versions of adult guns.

Length of stock

Many older guns will have short stocks. This is because people are generally bigger and taller than they were 50 or 100 years ago. Finding a 12-bore with a stock length of 13" or 13½" is not uncommon. Many have had rubber butt pads added to lengthen them. It may be an option to buy such a gun and simply remove the long rubber pad and replace it with a shorter one, adding to it as the child or teenager grows in stature.

Syndicate shooting

In the 'Golden Era' of driven shooting, Edwardian royalty, nobility, gentlemen of means and the wealthy, socially-ascendant beneficiaries of the industrial revolution entertained one another with lavish country house parties based around a lot of driven shooting. Large bags were the order of the day and the pace was kept up quite nicely until the Great War of 1914-1918 put an end to the kind of British society that was able to sustain this level of opulent social ritual.

The history of Edwardian shoots is well documented, and interested readers will find an excellent account in J.G. Ruffer's book *The Big Shots*, from which the following passage explains something of the scale on which formal shooting was organized: '*No expense was spared. The scale of the major shoots was staggering. It even caused some bankruptcies, Lord Walsingham and Maharajah Duleep Singh among them, but in an age when income tax was 5d. in the pound, the resources of the very rich could withstand extraordinary demands.*'

Shooting continued between the wars, although not on quite the scale of the early years of the century but the Second World War and the austere years that followed saw many of the great estates broken up and the grand houses demolished. Shooting continued among the landed classes but a new order became established in the countryside – and the shooting syndicate began to blossom.

Syndicate shooting opened shooting land up to those able and willing to pay rather than the old order of social exclusion that required invitations. If you were not 'the right sort' or were not connected, you did not get invited to formal shoots. With the advent of the syndicate, groups of like-minded shooting men could gather together, pool their funds, and buy the rights to certain acreage of land, employ a gamekeeper and shoot according to their ability to pay.

The post-war stereotype: Maurice Tullock's illustration of the syndicate from *Letters to Young Shooters* by Uncle Ralph.

Predictably, the landed classes initially looked down on the syndicate, which had a reputation of being made up of spivs, tradesmen who had made their fortune, City tycoons and others not quite of the character and breeding one would want to mix with. Douglas Middleton, writing in 1961 notes '*I well remember as a schoolboy the horror and disgust with which an old shooting friend informed me that 'the Colonel has let his shooting to a sinticate' and I must admit I still wish somebody would think of a rather less mercenary-sounding term for a gathering of sportsmen*'

However, the rise of the syndicate brought driven shooting within reach of a broader populous and no doubt helped fuel the current level of interest in the sport across the social divide. How many of us were first introduced to formal shooting via a syndicate? Possibly as a beater, or as the guest of an uncle or grandfather. Without the syndicate, those of us without land of our own would have found access to driven shooting almost impossible.

The sneering of the aristocracy gradually faded as the syndicate became embraced as the norm and at the beginning of the 21st century, shooting syndicates are now the established order. They may be large or small, expensive or (sometimes unbelievably) cheap, extremely well organized or so informal that 'organization' is a word one could barely use to describe their activities at all.

They all offer their members warm camaraderie and relaxation; the fruits of the hard work put in during the spring and summer can be reaped in the cold winter months and the stories told round an open fire or a wood-burner in the gunroom at the end of the day help to bond the members ever closer. The syndicate is made up of friends with equal shares and equal responsibility for making it work. The atmosphere in a good syndicate is friendly and good-humoured with plenty of teasing and long running jokes among old friends.

The syndicate also offers a thread of sporting continuity; new members join as old ones retire from the shooting line and the new members will be initiated into a club with rules of good sportsmanship and high safety standards, which they will again pass on in time. Children come along as beaters and dogs get exercise. The farmers have good reason to preserve the coverts and margins, develop the ponds and wetlands to attract waterfowl; and wildlife as a whole benefits from the existence of the shoot. The syndicate is often made up from the local community and can act as a binding agent, pulling people closer together.

With the syndicate established as the norm, the shooting world is now faced with yet another development: so-called 'corporate shooting'. This began with syndicates selling an occasional day to outsiders to help balance the books – a few thousand pounds extra in the bank account helped finance the rearing, feeding and other costs of running a shoot.

The idea of doing this on a purely commercial basis soon became apparent and shoots began to emerge where, instead of a syndicate buying the shooting rights for their own pleasure, an individual or company would rent them. The organiser would rear and release the birds, plan the shooting diary and sell each available day to a team of guns, usually at a fixed price per bird killed.

Commercial shoots allow a company to book a day's pheasant shooting as 'corporate entertainment'; an option often exercised by City firms, which has led to some controversy in traditional shooting circles. This is due to the perception that participants are only interested in killing large bags of pheasants, that they are not interested in the quarry, the produce or the spirit of shooting. Rather, the day is seen as an expensive social jolly featuring a lot of feathered clay-pigeons. In short, they are not *sportsmen* in the true sense of the word and they bring the sport of shooting and the whole way of life connected with it into disrepute.

Others argue that the money brought into shooting and the rural economy is essential and beneficial to a wide range of people and that detractors of commercial shoots are exercising the same snobbery once reserved for the shooting syndicate. Times change, they maintain, and we must move with them.

One benefit that commercial shooting has brought about is the access to once-exclusive shooting to anybody willing and able to pay. In the past such places were the preserve of the privileged few and were unobtainable to the masses. Nowadays, if you have the inclination, you can save up the three thousand pounds you will need to join a line of six guns shooting six hundred birds at Six-Mile-Bottom in Cambridgeshire and enjoy the sport of royalty.

On a less ambitious front, friends can arrange to shoot in six different counties on six different days instead of always shooting the same ground. The 'Roving Syndicate' is a recent evolution of the shooting syndicate. The friends form a line of Guns and book a day on a different commercial shoot each time they wish to shoot together. They then travel, stay in an hotel and enjoy good shooting in good company whilst maintaining the variety and interest that such travels with a gun can provide.

The Internet, that very modern phenomenon, has given rise to ever greater and faster dissemination of information and a number of enterprises have emerged to cater to the needs of shoots and shooters.

'The most important part of the day' Shoot Captain Peter Jones briefs the Guns on the rules before the first drive.

Shooting4all.com is a good example of a business that exists to link those with shooting for sale with shooters seeking a day somewhere different. If you shoot in a regular syndicate in Yorkshire but fancy a day's driven woodcock shooting in Wales, here you can find it. If you suddenly find you have a free Saturday and want to take your business partner for a day on the partridges, you will find what you need at a touch of the 'mouse'. It works for the shoots as well. If you have a cancellation or a last minute drop out and need to make up the numbers in your syndicate, you can advertise it and get a new member quickly.

All these developments show how shooting is moving into the new century and embracing the best of the new whilst maintaining the traditional values of sportsmanship and the love of the natural world. Let us hope that the easy access to shooting that our modern times bring does not harm these traditions. It is a sad fact that what is easy to come by is usually less respected or revered. It is hard to see how one can have as much involvement and respect when turning up to shoot a given quantity of birds in a locale you have no association with as when you have been an integral part of their rearing, feeding, habitat provision and protection.

Time will tell where our sport goes but I have every hope that it will continue to draw people into the countryside and help to embrace them with a better love and understanding of the environment and the cycles of life in which the shooting man becomes an integral part.

Douglas Middleton's forward to *Syndicate Shooting* in 1961 reads:

'One of the most significant changes has been the tax-forced decline of the big country-bred landowners and the transfer of their responsibility for game conservation to others who can better afford to spend the money on shooting: the tenants, individuals or syndicates, who now control so much of the shooting on our farms and feudal estates. I use the word responsibility deliberately, because I consider game to be something more than the object of an afternoon's pot hunting. It is an economic natural resource, a supplementary crop to farming and forestry, a source of trade and rural employment, but – above all – a natural heritage without which the British countryside and the culture it evolves would be vastly the poorer. Much of this responsibility now rests with the syndicates and they must be prepared to accept it and understand how to use it.'

His sentiments can be applied today to commercial shoots and syndicates alike.

Despite continuous improvement, the method of firing shot from a gun barrel via a centre-fire cartridge is essentially the same as it was in 1861.

AMMUNITION FOR VINTAGE GUNS

Cartridges

Cartridges used now are not radically different from those used by our grandfathers. Quality of powder and shot and methods of loading have improved and the cost of ammunition is lower now than it was when I was a teenager.

Most nitro-proved vintage guns are chambered for the 2½" cartridge and will happily accept modern ammunition. While there are many different brands and types of cartridge marketed, they fall into two major categories; those with fibre wads and those with plastic wads.

If you have a vintage gun that has been lapped out to remove pitting, it may be large in the bore. In this case, plastic shot cup cartridges can help eliminate the risk of 'balling'. This is when shot is fused together by the gasses escaping around the wad. This is more likely to happen in barrels of wide internal dimensions, as the wad may not expand sufficiently to block their passage. Those shooters who worry about the littering problem posed by the use of plastic shot cups may seek out products with 'photo-degradable' properties such as the excellent Hull Cartridge 'High Pheasant' brand.

I strongly believe that many shooters over-load their guns to make up for lack of skill or judgement. Invited to shoot in Italy recently, we walked up pheasants and were given 35g loads of No.6 shot to shoot them with. I thought this excessive until a team of American guests turned up at North Mymms to shoot driven partridges armed with 36g Eley VIPs!

English game guns behave best when appropriately loaded. One ounce of shot is perfectly sufficient for producing clean kills at game at ranges up to forty yards. No more is necessary or desirable. Lighter loads are also more forgiving to shoot in light guns. After years of experimentation, I now almost always use fibre wad 65mm (2½") Game Bore 'Pure Gold' cartridges for game shooting with a 12-bore. I use No.7 shot on driven partridges and No.6 for everything else. If I fail to kill my bird, rest assured, the cartridge is not to blame!

Lead shot

Will lead shot be available in the UK for any kind of shooting in ten years time?

The answer is unclear; some shooters anecdotally have almost accepted the demise of lead as inevitable. Others

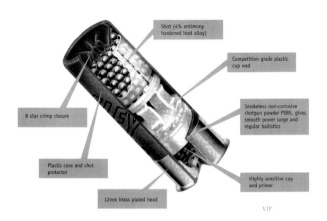

A modern plastic-shot cup cartridge; a clay load by Eley Hawk. Plastic shot cups became popular in the 1960s and are in widespread use today. Many game shoots discourage their use due to the littering of plastic wads left on the ground.

believe it will be allowed to remain in use for game and clays. Whatever the outcome, the fight is on. It has started already and some countries have lost already. European legislation is big wheels turning and once the wheels start, they are loath to stop. They gain a momentum and then the end becomes inevitable, as the movement becomes more important than the issue or its merits.

Lead is currently banned in the UK for the shooting of any waterfowl or for use over wetlands. This has led to cartridge manufacturers exploring other options. None are as effective as lead for all-round performance, all are more expensive and many are unsuitable for using in vintage guns.

Lead substitutes in vintage guns

Steel shot (actually iron) should not be used in vintage guns in my opinion. Should you insist on using it, do not fire it through a gun proved more tightly than ½ choke and do not use it in Damascus barrels. Many steel-loaded cartridges require special CIP proof for steel shot. Do not ignore the dangers of using this material in old guns. They were not designed for it and will not respond well to its use. Steel shot is lighter than lead and does not kill as cleanly, leading to more wounded birds. It is cheap but has little else in its favour.

Hevi Shot is widely touted as a substitute for lead

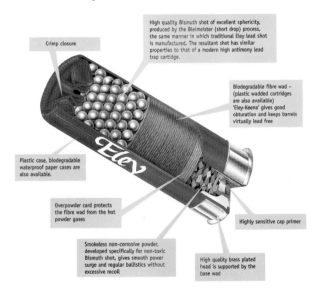

A modern fibre-wad cartridge; a bismuth load by Eley Hawk. This type of cartridge has been in use since Daw introduced the centre-fire in the early 1860s.

shot. Ballistic performance and field reports are impressive but at $6 per shot it is beyond the reach of most shooters and is too hard to use in vintage guns.

Tin is light and ineffective as shotgun ammunition. I predict that after the development of lead-load replacements has passed the current transitional period, that tin will cease to be offered.

'Hevi-Shot' is heavier than lead and has good potential as shotgun ammunition, though it is harder than steel and will not be kind to your fine old barrels. It is also extremely expensive in comparison with all other alternatives. It currently suffers from the difficulty faced in getting it to form uniform-sized, round shot; though by all accounts it patterns very well.

Bismuth is the most effective lead substitute and is now usually mixed with a small percentage of tin to combat its natural brittleness. It is a little lighter than lead but patterns slightly more evenly by way of compensation. It is suitable for use in any gun proved for the equivalent load of lead and produces good results in the field. It is currently around four times as expensive as the equivalent lead loaded cartridge.

For shooting inland wildfowl, including Canada geese, with my vintage game guns, I find the Eley Maximum cartridge loaded with 34g of No.3 or No.4 Bismuth is effective at normal ranges, and gentle on the guns in both recoil and barrel abrasion. It can be fired with confidence through choked barrels, exactly as lead.

It is certain that progress will continue to be made and new offerings in 'non-toxic' ammunition appear each new season. Hopefully the prices will come down in time. However, lead remains the most effective ammunition for use in old guns and long may we continue to have use of it.

Do vintage guns have a future?

In the years during which this book has gone through the process of conception, writing, editing and, finally, printing, I have detected a real stir in the market. The magazine articles of David Baker, Mike Yardley, Donald Dallas, Chris Austyn and myself on the subject in the British shooting press, and high quality in-depth contributions to the American quarterly *The Double Gun Journal* by numerous writers, have raised the qualities of the pre-war British shotgun in the consciousness of collectors and shooters alike.

A number of maker-specific histories of the famous firms have been published. Where there is information, enthusiasm often follows and interest heats up in a particular maker. This has an effect on the market. Never before has the owner of a gun been able to gather information about it and those who made it more readily.

Prices for hammer guns have risen steadily and I detect a greater level of sophistication beginning to emerge in the gun-buying public as enthusiasts look towards the weird and wonderful and become a little more adventurous in their choice of weapon. All this is encouraging.

We in the gun world must however, remain vigilant, for there are those who would deny us our sport and the objects of our enthusiasm. European legislation to standardize everything leaves little space for minority exceptions and ever more stringent proof practices led by CIP legislation, greater restrictions on the age at which people can handle a weapon of any kind, food handling restrictions, changes to the General Licenses granted for vermin shooting and ever tighter security demands all put pressure on shooting sports and gun ownership.

We shooters of antique shotguns are of no value to the law makers and civil servants. If we are passive, we will be legislated out of existence. If you think it could never happen, ask a British pistol shooter.

I urge all readers to seek awareness of the politics surrounding shooting and gun ownership and provide as united a front as we can; for the vocal but aggressive minority who would ban all private firearms ownership and outlaw the killing of any wild animal are a lunatic fringe; but a disproportionately influential one.

When you next stand on a flight-line with your Grant hammer gun, wait in the woods for oncoming boar with your Woodward .450/400 black powder express, sit in the hide with your Greener 'Facile Princeps' or stand at the peg with your Purdey side-lock, remember that you are continuing a timeline of man and gun that stretches back to before Hawker's time. It is your privilege, your heritage and your right. Value it and defend it.

USEFUL ADDRESSES

I have included only a small number of useful contacts whose work I can either vouch for myself, or who come recommended by those I know and trust.

AUCTION HOUSES

Sotheby's
1334 York Avenue at 72nd St
New York,
New York 10021
Tel. 212 606 7000
www.sothebys.com

Charlton Hall
912 Gervais Street
Columbia
South Carolina
29210
Tel: 803 779 5678
www.charltonhallauctions.com

J.C. Devine Inc
Auctioneers/Appraisers
PO Box 413 - 20 South Street
Milford, NH 03055
Tel: (603) 673-4967
info@jcdevine.com

Bonhams & Butterfields
595 Madison Avenue, 6th Floor
New York, New York 10022
+ 1 212 644 9001

James D. Julia
PO Box 830
Fairfield, ME 04937
Tel.207-453-7125
e-mail: juliagun@juliaauctions.com
www.juliaauctions.com

Holt & Company
The Gun Room, the Sandringham Estate
Church Farm Barns, Wolferton
Norfolk PE31 6HA
www.holtandcompany.co.uk

Sotheby's (UK)
34-35 New Bond Street
London W1A 2AA
www.sothebys.com

Christie's
85 Old Brompton Road
London SW7 3LD
www.christies.com

Bonham's
101 New Bond Street, London W1S 1SR
www.bonhams.com

Scotarms Ltd
The White House, Primrose Hill
Besthorpe, Newark NG23 7HR
www.scotarms.co.uk

Gavin Gardiner Ltd
Hardham Mill Business Park
Mill Lane
Pulborough
West Sussex RH20 1LA
www.gavingardiner.com

Southams
8 Market Place
Oundle, Northamptonshire PE8 4BQ
www.southams.com

GUNMAKERS

James Purdey & Sons
Audley House, South Audley Street
London W1
www.purdey.com

Boss & Co Ltd
16 Mount Street
London W1K 2RH
Telephone +44(0) 207 493 1127
Fax +44 (0) 207 493 0711
www.bossguns.com

J. Roberts & Son
22 Wyvil Road
London SW8 2TG
www.jroberts-gunmakers.co.uk

Holland & Holland
31-33 Bruton Street, London W1J 6HH
www.hollandandholland.com

Atkin, Grant & Lang
Broomhill Leys, Windmill Road
Markyate, Hertfordshire AL3 8LP
www.atkingrantandlang.co.uk

William Evans
67a St James's Street, London SW1A 1PH
www.williamevans.com
Telephone +44 (0)20 7493 0415
Fax +44 (0)20 7499 1912

Holloway & Naughton
Turners Barn, Kibworth Road
Three Gates, Illston-on-the-Hill
Leicestershire LE7 9EQ
www.hollowaynaughton.co.uk

Dickson & MacNaughton
21 Frederick Street
Edinburgh EH2 2NE
Tel: 0131 225 4218
www.dicksonandmacnaughton.com

W W Greener (Sporting Guns) Limited
Stoppers Hill, Brinkworth
Chippenham, Wiltshire SN15 5AW
Telephone +44-1666-510351
Fax +44-1666-510948
www.wwgreener.com

Westley Richards & Co. Ltd
40 Grange Road, Bournbrook
Birmingham B29 6AR
Telephone +44 121 472 2953
Fax +44 121 414 1138
www.westleyrichards.com

E.J. Churchill Group Ltd
Head Office: Park Lane, Lane End
High Wycombe, Bucks HP14 3NS
Tel: +44(0)1494 539202
Email: gunmakers@ejchurchill.com
www.ejchurchill.com

Cogswell & Harrison
Thatcham House, 95 Sussex Place
Slough , Berkshire SL1 1NN
Telephone: +44 01753 520866
E-mail: info@cogswell.co.uk
www.cogswell.co.uk

William Powell & Sons
35–37 Carrs Lane
Birmingham B4 7SX
Tel: +44 (0)121 643 0689/8362
www.william-powell.co.uk

William James Grant
2 Old Brompton Road
London SW7 3DQ
Tel: (00 35) 7224 92049
Mobile: (00 35) 7996 97477
Fax: (00 35) 7225 18667
www.williamjamesgrantguns.com

SHOOTING TUITION

Michael Yardley
Witham, Essex
Telephone 07860 401068
yardleypen@aol.com

The West London Shooting School
Sharvel Lane , West End Road
Northolt, Middlesex UB5 6RA

MUZZLE LOADING AND BLACK POWDER SUPPLIES

Peter Dyson
3 Cuckoo Lane, Honley, Holmfirth
Yorkshire HD9 6AS

BARREL WORK

Briley Manufacturing, Inc.
1230 Lumpkin Road
Houston,
Texas 77043
E-mail: SMP@Briley.com

F. J. Wiseman & Co Ltd
262 Walsall Road
Bridgtown, Cannock WS11 3JL
01543 504088

ENGRAVING

Barry Lee Hands
Engraving Studios
174 Fernview Lane
Bigfork, MT 59911
Tel: (406) 249-4334
E-mail: Barry_hands@yahoo.com

REPAIRS & SERVICING

Kirk Merrington
207 Sierra Road
Kerrville
Texas 7802
kirknkk@ktc.com

Keith Kearcher
60232 Ridgeview Dr. East
Bend, OR 97702
Tel: (541)-617-9299

James Tucker
P.O Box 366, Medford
Oregon 97501
Tel: 1-541-245-3887

Darlington Gun Works Inc
516 S Governor Williams Hwy
Darlington,
South Carolina 29532
Tel: (843) 393-3931

David Mitchell
Ickenham, Middlesex
dmitchellgunmaker@supanet.com
01895 633152

David Sinnerton
26 Tennison Close, Horsham
West Sussex RH12 5PN

SLEEVING

Teague Engineering (Nigel Teague)
Edinburgh Way, Leafield Industrial Estate
Corsham, Wiltshire SN13 9XZ

GUN FITTING

Glen Baker
Thomas Bland
Woodcock Hill
Telephone: (570) 864 3242
bland@epix.net

STOCK ALTERATIONS

Trevallion Gunstocks
9 Old Mountain Rd.
Cape Neddick
ME 03902
Tel: 207-361-1130

Jim Greenwood
6400 SW Hunter road
Augusta
Kansas 67010

Jim Spalding
112 East Road, West Mersea
Essex CO5 8SA
Telephone (01206) 382477

BESPOKE AMMUNITION

Ballistic Products UK (Alan Myers)
Vicarage Farm, New House Lane
Winmarleigh, Garstang
Lancashire PR3 0JT

SHOOTING BAGS AND CASES

Parsons & Sons
Unit 4, Block 6
Shenstone Trading Estate
Bromsgrove Road
West Midlands B63 3XB

Brady of Halesowen
N8 International (suppliers)
232 Ferme Park Road
London N8 9BN

IMPORT / EXPORT

The Cheshire Gun Room
29 Buxton Road
Heaviley, Stockport SK2 6LS
www.cheshiregunroom.com

LOCATING MISSING GUNS

Matched Pairs Ltd
Midgeholme, Moredon
Sedgefield, County Durham TS21 2EY
www.matchedpairs.com

FINDING EXAMPLES OF RARE PATENTS

Pantiles Vintage Guns
8 Union Square
The Pantiles
Tunbridge Wells , Kent TN4 8HE

ASSOCIATIONS

The Vintagers
Ray Poudrier
P.O.Box 31
Hawley, MA 01339
USA
Tel.413-339-5347
www.vintagers.org

The Worshipful Company of Gunmakers
The Proof House
48/50 Commercial Road
London E1 1LP
Tel: (020) 7481 2695
Member of: Game Conservancy

The Muzzle Loaders Association of Great Britain
MLAGB Membership Office
82a High Street, Sawston
Cambridge CB2 4HJ
Email: membership@mlagb.com
Telephone 01223 830665
Fax 01223 839804

The Gun Trade Association Ltd
PO Box 43
Tewksbury Gloucestershire GL20 5ZE
Tel: 01684 291 868
www.guntradeassociation.com

British Association of Shooting & Conservation (BASC)
Marford Mill, Rossett
Wrexham LL12 0HL
www.basc.org.uk

VALUATIONS/ADVICE ON BUYING AT AUCTION

Vintage Guns
232 Ferme Park Road, London N8 9BN
www.vintageguns.co.uk

STOCK BLANKS

Steve Turgay Sidki
11 Rowan Close
Meophan Gravesend DA13 oEJ
www.gunstockblanks.co.uk

BIBLIOGRAPHY

British Gunmakers Vol. One (London), Nigel Brown   Quiller

The Shooting Field, Holland & Holland, Quiller

Purdey: The Definitive History, Donald Dallas, Quiller

Birmingham Gunmakers, Douglas Tate, Safari Press

Modern Sporting Guns, Christopher Austyn, Sportsman's Press

Letters to Young Shooters, Uncle Ralph, Rich & Cowan

Purdey's: the Guns & the Family, R. Beaumont, David & Charles

Boothroyd on British Shotguns, G. Boothroyd, Sand Lake Press

Gough Thomas's Gun Book, G.T Garwood, A&C Black

Syndicate Shooting, J.F Standfield, Herbert Jenkins

Shotgun Marksmanship, Stanbury & Carlisle, Herbert Jenkins

The Art of Shooting, Charles Lancaster, Ashford Press

Shotguns & Cartridges for Game & Clays, G.T. Garwood, A & C Black

The Shooting Man's Bedside Book, 'BB', Merlin Unwin Books

Game Guns & Rifles, Richard Akehurst, Arms & Armour Press

Sidelocks & Boxlocks, Geoffrey Boothroyd, Safari Press

Amateur Gunsmithing, Mills & Barnes, Boydell

The British Shotgun Vol 1, Crudgington & Baker

The British Shotgun Vol 2, Crudgington & Baker, ABE

Holland & Holland The Royal Gunmaker, D. Dallas, Quiller

The Early Purdeys, L. Patrick Unsworth, Christie's Books

Heyday of the Shotgun, David Baker, Swan Hill

The Big Shots, Jonathan Ruffer, Quiller

Sporting Shotgun, Robin Marshall-Ball, Saiga Publishing

The Shotgun Handbook, Mike George Crowood

The Shotgun, MacDonald Hastings, David & Charles

Shotguns & Gunsmiths, Geoffrey Boothroyd, Safari Press

Book of Shotguns, James Marchington, Pelham Books

The Shotgun, Michael Yardley, The Sportsman's Press

The Modern Shotgun, Vols 1,2&3, Sir G. Burrard, Herbert Jenkins

The Gun & its Development, W.W. Greener, New Orchard Edns

Modern Shotguns, W.W. Greener, Tideline Books

In The Gun Room, Sir Gerald Burrard, Herbert Jenkins

The Shotgun, T.D.S. & J.A. Purdey, A&C Black

Pigeon Shooting, Richard Arnold, Faber & Faber

The Shotgun History & Development, G. Boothroyd, Safari Press

Lock, Stock & Barrel, Adams & Braden, Safari Press

The Sport of Shooting, Owen Jones, Arnold

The Shooter's Handbook, Richard Arnold, MacDonald & Evans

Automatic & Repeating Shotguns, R. Arnold, Nicholas Kaye

The Diary of Colonel Peter Hawker, Peter Hawker, Bath Press

High Pheasants in Theory & Practice, Sir R. Payne-Gallwey, Tideline Publishing Co.

Experts on Guns & Shooting, G.T. Teasdale-Buckell, Ashford Press

Modern Sporting Gunnery, Henry Sharp

History of W&C Scott, Crawford & Whatley, Rowland Ward

Pigeon Shooting, 'Blue Rock', Gunnerman Press

Boss & Co: Makers of Best Guns Only, Donald Dallas, Quiller

The House of Churchill, Don Masters, Quiller

Cogswell & Harrison, Cooley & Newton, Sportsmans Press

Rifle & Gun, L.B. Eskrit, MacDonald & Evans

Shooting & Gunfitting, Arthur Hearn, Herbert Jenkins

Clay Pigeon Shooting: a History, Michael Yardley, Blaze

Instructions to Young Sportsmen, Col. P. Hawker, Field Library

Game & Gun Magazine 1924–1936

Modern Breechloaders 1871, W.W. Greener

GLOSSARY OF SHOOTING TERMS

ANSON PUSHROD
Or 'push-button forend catch'. This method of attaching the forend to the barrel loop by means of a sprung rod, released by pressing its extremity, quickly became the favoured forend catch for side-by-side shotguns.

ANSON & DEELEY ACTION
This hammerless, barrel-cocking action, patented in 1875, has become generically known as the 'boxlock'. It was the first really successful hammerless gun and millions have been made.

ARTICULATED TRIGGER
The front trigger on some double shotguns is reverse-hinged in order to prevent bruising of the trigger finger upon firing the second barrel.

ARCADED FENCES
A style of fence, shaped as if it had forward-facing arches encircling the bore.

ACTION BAR
A breech loading shotgun has a bar of metal protruding from the body of the action upon which the barrels rest when the gun is closed. This is known as the 'bar'.

ACTION
The action is the main metal part of the gun which connects the stock and the barrels.

ACTION FLATS
The flat part of the bar, facing upwards, which opposes the barrel flats when the gun is closed, also called the 'Table'.

ACTION FACE
The vertical face of the action of a breech-loader, which closes the end of the chambers and forms a barrier when the gun is closed. Also called the 'Standing Breech' or 'Breech Face'.

ACTIONER
The name given to a gunsmith whose job it is to file the gun's action from a rough forging.

APPRENTICE
British gunmakers traditionally served a seven-year apprenticeship with a master gun maker before being considered skilled craftsmen. Early apprenticeships involved hardship and the apprentice was housed and clothed by his master but received no wages. An apprentice usually learned one of the gunmaking crafts: for example Stocking or Finishing and when his time was served he would become a stocker or a finisher.

ALKANET ROOT
Alkanet (*Anchusa officinalis, Anchusa tincturia*) is a European weed used to make 'Red Oil', which is applied to a stock before the finish, in order to deepen the color and contrast. Some gunmakers simply add raw linseed oil to the dried root and let it steep for months, others also use turpentine.

BALL & SHOT GUN
Early smooth-bored double guns with fold-down leaf sights and open chokes and could be loaded with a solid ball or shot shell, as required. These developed into shotguns with rifled chokes like Holland & Holland's 'Paradox' or Westley Richards's 'Fauneta', both of which fired a special conical bullet but still performed quite well as shotguns.

BALL FENCES
Fences with a distinctly rounded profile, providing the depth, and therefore strength, that was arguably lacking in earlier pinfire-type fences.

BAR-ACTION
The term used for a gun in which the mainspring of the locks is housed in a recess cut into the action bar, in front of the hammer or tumbler. This has become the preferred style for 'best' guns.

BARREL COCKING
A hammerless gun which is automatically cocked by the fall of the barrels acting on cocking dogs, which in turn re-set the tumblers is known as a 'barrel-cocking action'. The first successful design on this principal is the Anson & Deeley boxlock.

BACK-ACTION
The term is used for a gun in which the mainspring of the locks is fitted in a recess cut into the stock, behind the hammer or tumbler.

BAR-IN-WOOD
A style of gun in which all or part of the steel action bar is covered by wood. It was commonly employed as a style in the early days of breech-loaders as it was aesthetically similar to the familiar muzzle loader. It continued to be popular until the early 20th century.

BORE
The internal tube of a shotgun barrel. The size of the bore is determined by the number of equal sized balls that could be made form a pound of pure lead. Therefore, a 12-bore is the diameter of a lead ball weighing one twelfth of a pound.

BOLT ACTION
A well-used system of closing and opening the breech by means of a manually operated sliding-bolt. Best known as a rifle action, like the Mauser, it has been used periodically on low-priced shotguns like Webley & Scott's 1960s .410.

BEAD
A shotgun traditionally has a small bead at the muzzle-end of the rib to aid target acquisition. It is usually made of brass, but can be ivory. Some trap guns have an intermediate bead mid-way down the rib.

BEST GUN
The term 'best' applied to a British gun indicates that it has been made to the highest standard regardless of cost.

'B' QUALITY
Many gunmakers made guns of various qualities to suit the pockets of their customers. The term 'B Quality' usually meant a gun of very high quality but with less lavish engraving and finish than a 'best' gun.

BARREL
The barrel is the part of the gun through which the projectile travels. It consists of (internally) chamber, forcing cone, bore, choke and muzzle. Externally it carries the ribs, the bead, the extractors and the lumps.

BREECH
The open end of the barrels that houses the chambers and into which cartridges are placed (hence 'breech-loader'). Muzzle-loaders have breech plugs, which are securely screwed into the breeches to block them off.

BREECH LOADER
A gun loaded from the muzzles with loose powder, wadding and shot, by means of a ram-rod.

BREECH PLUG
The solid metal screw-plugs that seal the breech ends of a muzzle-loader

BREECH PIN
The biggest, strongest screw that holds the action (via the top strap) to the trigger plate and braces the stock-head in position.

BRIDLE
The metal plate inside a lock, which retains all the moving parts of the lock-work on a side-lock gun.

BENT
The bent is a notch cut into a tumbler, into which a sear fits to hold it in place until fired.

BELGIAN DAMASCUS
Belgium was a prodigious manufacturer of Damascus barrel tubes for the gun trade. Many British and American makers used Belgian tubes. They were held to be of very fine figure but inferior to best English Damascus for strength and durability.

BROWNING
Damascus barrels are traditionally given a brown finish, which enhances the color contrast of the iron and steel components. It is a rusting process, which is both skilled and messy.

BLACKING
The furniture of a gun is traditionally blacked, as are steel barrels. Hot blacking is considered better and harder wearing than cold blacking.

BLUEING
The terms blueing and blacking are interchangeable.

BIRMINGHAM PROOF
There are two proof houses in the UK: London and Birmingham. They operate in the same manner and on the same terms but guns proofed in either house will carry different marks.

BUTT SOLE
The flat end of a gunstock, which rests in the shoulder pocket of the shooter when the gun is fired.

BUTT PLATE
A plate screwed onto the butt. It can be metal, horn, wood or plastic.

CHAMBER
A straight, machined recess in the breech end of a barrel, made to accept a cartridge. The chamber is cut to the length of cartridge case for which the gun is designed.

CHAMBER CONE
Also called a 'forcing cone', this is a graduated slope from the end of the chamber into the bore of the gun.

CAPE GUN
A gun built as a double shotgun but with one barrel rifled. It was intended for those travelling to Africa or India who wanted a gun with which to shoot both birds and dangerous game.

CAST ON
The bend given to a stock to make it point to the left.

CAST OFF
The bend given to a stock to make it point to the right.

'C' QUALITY
Some 'best' gunmakers offered lower quality guns, often made in Birmingham and stamped them in lower grades.

CENTRAL VISION
Most shooters have one eye which is dominant in picking a sight picture, called the 'master eye'. Most right-handed shooters have a right master eye. However, some shooters have a master eye which is not fully dominant. They are said to have 'Central Vision'.

Such right-handed shooters will require more 'cast-off' than those with a right master eye.

CROSS-OVER STOCK
A severely bent stock, enabling a left-eye dominant shooter to shoot from the right shoulder, or vice versa.

CHAMBERLESS GUN
'Chamberless guns' were actually chambered to take thin brass cartridges. They were popular at the turn of the last century for wildfowling guns but never gained general adoption.

CROLLE
Crolle is a term used for a fine Belgian Damascus barrel tube.

CIRCLE
The circle is the concave surface of a barrel lump.

CHOPPERLUMP BARRELS
When steel barrels were developed (1890s), a system was adopted for best quality barrels by which the forging for each tube included the lump as well as the tube. The two barrels were then joined together centrally but there was no need to make a second joint to fit the lumps, as they were integral. This removed one more potential area of weakness.

CENTRE-FIRE
Developed in 1861 by Daw of Threadneedle Street, London. The centre-fire cartridge as we know it today superseded pinfire and has changed little since. A suitably proofed 1861 Daw system hammer gun will happily fire modern shotgun ammunition.

COMB
The comb is the part of a stock on which the shooter's cheek rests when levelling the gun at a target.

CARVED FENCES
Different makers and different eras are often marked by the carving on the fences of the action. Common motifs are oak leaves, acorns, fleur-de-lys, ribbons, arcs and acanthus leaves.

CHOKE
Choke is the constriction at the muzzle of a shotgun that affects the spread of shot. A greater constriction will concentrate the shot to a greater extent, thereby extending the effective killing range of a given load.

CHURCHILL RIB
This is a raised, filed, tapered rib designed by Robert Churchill to complement his short 'XXV' barrelled guns. It has become generally adopted for this purpose by other makers.

COIL SPRING
Coil springs are not widely used in British gunmaking for powering the locks. Exceptions are the Ward 'Target Gun', the T. Woodward spiral-spring

'Acme' and other similar patents. The locks of English guns will generally be fitted with leaf springs.

COCKING INDICATOR
In the early days of hammerless guns, shooters (who habitually carried their hammer guns closed and at 'half-cock') were suspicious of the new-style guns that were automatically cocked and therefore dangerous when closed. Cocking indicators showed whether the tumblers were 'at cock' or at rest.

CROSS-OVER STOCK
A stock designed to enable a right-handed shooter to shoot from the right eye, using his left master eye and shooting with both eyes open. Cross-over stocks can also be made in reverse for left handed shooters with right master eyes.

DROP-LOCK
The term 'droplock' has been coined by some writers to describe the 1897 Deeley & Taylor patent 'hand-detachable boxlock' made famous by Westley Richards. It is not a helpful term, since the locks are certainly not intended for 'dropping'.

DROP POINTS
The panel on the side of a shotgun where the wood joins the action is often shaped into a tear-drop at the point where it meets the hand of the stock. This is often omitted in lower-quality guns.

DAMASCUS BARREL
'Damascus' is a term that has become generic in describing barrel tubes that were hand, or machine, hammer-welded from twisted rods of iron and steel. They vary in quality and original cost greatly but the best were the equal of steel barrels in performance and strength. The premier barrel-making centres were in Belgium and the English midlands.

DENT
When a shotgun barrel is knocked against a hard surface, it may become indented, causing a low spot on the exterior and a raised spot o the interior. Dents can be removed by skilled gunsmiths using a hydraulic raising tool and hammering the tube back into shape.

DIAMOND GRIP
Most shotgun stocks have a rounded shape to the 'hand'. Holland & Holland are known for shaping this into a soft 'diamond' shape for better grip. Some other makers have copied this style over the years.

DEELEY & EDGE FASTENER
This is a forend-release fastener consisting of a plate inlet into the forend surface with a lever in it. The lever is pulled down to release the forend from the barrel loop. It is

common on British shotguns, but not as common as the Anson push-rod.

EJECTOR
Before 1874 double guns had to be unloaded manually. Then, Needham produced the first workable (though rather ungainly) ejector system and his work was quickly developed by others. The most widely used systems now used are the so-called 'Southgate', which is an over-centre type ejector of great simplicity and reliability and the Boss 'coil spring' type. Both systems are housed in the forend.

EJECTOR GUIDE
The ejector guide is the split extractor rod, which protrudes from the breech end of the barrels when a gun is opened. The quickest way to tell if a gun is an ejector or a non-ejector is to open the breech; a non-ejector has a single extractor rod, an ejector has a split pair of rods which can function independently of one another.

EJECTOR SPRING
All forend-housed ejectors are operated by springs. Some, like the Deeley and the Southgate use leaf springs, the Boss uses coil springs. Greener's ejector was powered by the mainspring.

EXTRACTOR
See 'Ejector Guide'. Extractors are one-piece and lift the cartridges clear of the breech for ease of removal.

EXPRESS
The term 'Express' was coined by Purdey to describe their powerful, flat trajectory ammunition and the rifles they designed to fire it. These are typically large-calibre double rifles designed for use on dangerous game. Originally designed as a black-powder cartridge, later 'nitro-express' ammunition is even more powerful.

EXPORT GUN
This is a term used derisively by Greener to describe the lowest quality gun intended for sale outside the UK. These are generally Belgian or British hammer guns of poor quality.

FORSYTH
Forsyth was a Scottish minister who was very influential in the development of the sporting gun. His ignition system led directly to percussion muzzle-loading and was quickly developed by others into self-contained cartridges. Rev. Forsyth employed James Purdey as Head Stocker until Purdey set-up on his own in 1814.

FORCING CONE
The forcing cone is the portion of the internal surface of a barrel tube, which leads from the chamber to the bore. It can be of various lengths and degrees of graduation.

FLUTED FENCES
A style of carved fence with a flattened,

engraved top, outside edge. The fluted fence is a 'house style' of Stephen Grant hammerless guns.

FINISHER
A specialist gunmaker whose jodb it is to assemble the parts of a shotgun that arrive from the stocker, barrel maker, lock maker, actioner and engraver. The finisher has to be both metal worker and woodworker and is perhaps the most widely skilled of the gunmaking specialists.

FURNITURE
The term used to describe the parts of a gun that are neither lock, stock nor barrel. Typically these include trigger guard, top-lever, hammers, etc.

FIELD TRIAL
The Field was the leading country sports magazine of the 19th Century. Its long-time editor Dr JH Walsh was a very powerful figure in Shooting and gunmaking circles and organized trials of Muzzle-loaders vs Breech-loaders and Choke-bored vs Cylinder-bored guns, which were very influential.

FALLING BLOCK
A type of rifle action in which the operation of an under-lever causes the block to depress below the entrance to the bore. This enables loading and unloading at the breech. The best known patents are the Martini and the Gibbs-Farquarson.

GAME GUN
A double gun designed for the shooting of (usually driven) game. This is generally a breech-loading, double ejector with a barrel choked open in the right barrel and tighter in the left, weighing between 6lb 2oz and 7lb in 12-bore. It will usually balance around the hinge pin and have a concave rib with single bead sight and barrels of 25"–30" in length.

GUNSMITH
The name given to a craftsman who has served an apprenticeship in the gun trade and learned one of the gun making disciplines.

GUN MAKER
The name of a person whose name appears on the guns he sells, and who is also a fully apprenticed and trained gunsmith. A barrel maker working for James Purdey would be considered a gunsmith but James Purdey himself would be considered a Gun Maker.

GUN
The British refer to a shotgun as a 'Gun', as opposed to a 'Rifle'.

GRAIN
A 'grain' is a measurement, often used to describe the weight of a bullet or a measure of shot.
'Grain' is also used to describe the fibres in the wood of a gun-stock.

Greener Cross Bolt
A circular-sectioned bolt that extends through the fences and, by way of a rib extension, best known when used in tandem with the double Purdey under-bolt and named the 'Treble Wedge Fast'. It was widely copied by other makers, especially for lower-grade guns but few, other than Greener, adopted it for 'best' guns.

Hammer
The external hammer is the limb, which falls to detonate the cap on a percussion muzzle-loader or to strike the pin on a pinfire or connect with the striker on a centre-fire hammer gun.

Hammer Gun
A gun with external hammers.

Hammerless Gun
A gun with internal hammers (then known as 'tumblers'). The first successful hammerless breechloader was patented in 1871 by T. Murcott , of London.

Heel
The extremity of the butt sole parallel to the comb

Horns
The section of the stock on a sidelock which extend from the grip to the fences. They are prone to damage, being very thin.

Hand Detachable Lock
Holland & Holland sidelocks often feature a lever on the outside of one lock-plate, which enables the locks to be removed without the use of a turn-screw. Deeley & Taylor's hand detachable boxlock features a hinged floor-plate, which can be opened without tools and the locks removed for safety or replacement.

Hook
The face of the front lump on a breech-loader which contacts the hinge pin and forms the pivot when the gun is opened and closed.

Horn Butt Plate
Some shotguns have a plate attached to the butt, this can be checkered, carved or lined. When horn is used as a material it is the horn of a buffalo. Best guns do not generally have butt plates.

Ironmonger
The purveyor of ironmongery: buckets, nails, ladders, pans, traps etc, which most small towns used to have. Ironmongers commonly sold guns. They would often buy these from Birmingham makers and would have their own name put on the barrels or locks. This is one reason why there appear to be so many gunmakers selling English guns in the 19th and early 20th centuries, although most of the guns were actually made by two or three big Birmingham firms like Scott, Webley or Bonehill.

Inert Action
Early breech-loading actions were 'inert' and required manual raising of the barrels to close the gun and subsequent manipulation of a lever to lock them in position. Good examples are the Jones patent of 1858 and Robert Adams's 1863 patent. Later systems employed a spring to return the lever and engage the locking bolt as soon as the barrels were closed. These are known as ''snap-actions''.

Jones Under Lever
The first totally reliable and widely adopted inert locking mechanism for breech-loader. It was developed by Birmingham gun maker Henry Jones in 1859. It has been used on every type of double rifle and shotgun and vintage examples still provide secure locking. Its main disadvantage is the relatively slow speed of operation in comparison with snap-actions.

Keeper's Gun
Many gunmakers, notably Westley Richards, Greener and Blanch, made guns of sound quality but plain appearance for use by gamekeepers and foresters who needed a gun for hard work at a modest price.

Lump
The downward-projecting metal forgings protruding from the barrel flats of a breech-loader. They are typically used as points on which the gun pivots when opened and by which the gun is locked, via a bolt engaging with the rear face when closed. Damascus barrels typically had dovetail lumps, as did second quality steel barrels. Best steel barrels are made on the 'chopper-lump' system.

Lapping
Lapping involves removing metal from the inside of a gun-barrel to remove pitting, to enlarge the bore or to install a recess choke.

Laminated Steel
Laminated steel is a type of 'Damascus' barrel, which was machine hammer-welded. It was made in England and, while very strong, was not as attractive as true Damascus.

Lefaucheaux Gun
The first successful breech-loading gun offered in Britain after the Great Exhibition in 1851. Joseph Lang was one of the first English gun makers to popularize the design and it proved that breech-loaders had a future and could truly rival muzzle-loaders. The original design was not sufficiently robust but it was quickly improved.

Lever Cocking
Breech-loading hammerless guns were typically cocked by either the fall of the barrels (like the Anson & Deeley) or by manual manipulation of an external lever (like the original MacNaughton 'Edinburgh Gun'). Many lever-cocked guns were very workable but the system gave way to barrel cocking (and to a lesser extent spring-cocking, like the Purdey) in popular application.

Lock
The mechanism used to fire a gun. It consists of a spring and a tumbler or hammer, which the spring powers to achieve detonation via a striker. Other parts of a lock are there for the purposes of holding the hammer and spring in place. In sidelocks, the locks are placed on lock plates and mounted either side of the gun. In boxlocks, they are located internally within the action and on trigger plate actions they are mounted on the trigger plate.

Long Sufferers' Association
The association of fully apprenticed London gunmakers.

Lift-up Top Lever
Westley Richards patented a top-lever, with a bolt emerging from the face of the action to engage a bite in the barrels. This was operated by pushing the top-lever upwards.

Lock Pin
The long screw (or 'pin') which enters one lock plate and screws in to the other lock plate, having passed through the stock. It secures the locks to the stock and action.

Lock Plate
The externally visible metal plate on which the internal lock work of a sidelock is mounted.

Lock Maker
The specialist firm, or craftsman, who produced the locks for larger gunmakers. Most London and Birmingham firms bought their locks from firms like Brazier and Stanton, who were based in the midlands, mainly Staffordshire.

Leaf Spring
A spring made from folded steel formed into a 'V'. This is the usual spring used to power the locks in British gunmaking.

Leaf Sights
Hinged 'V' shaped sights mounted on the rib of a double rifle or cape gun and regulated to the distances indicated. Typically, they are arranged in sequence of range (say 100, 200, 300 and 400 yards) and can be flipped up once approximate range has been calculated.

London Proof
The London Proof House is the older of the two English proof houses. It stamps guns using its own unique set of proof stamps. These have changed over the years and can therefore be used to help discover where and when a gun was made.

Master
The apprenticeship system for gun-makers linked an apprentice to a master gun maker for seven years. The apprentice was fed, housed and clothed by his master and was trained in the gunmaking disciplines. Latterly, apprentices did not live with their masters but were trained by him and their output was considered his work.

Muzzle
The foremost extremity of a barrel, from which the shot or bullet exits the gun. A 'muzzle-loader is loaded by having powder, wadding and shot introduced via the muzzle.

Muzzle Loader
A muzzle-loader is any gun loaded with powder, wad and shot or ball, via the muzzles, using a ram-rod. Muzzle-loading was the norm from the flint-lock era until the invention of the pin-fire cartridge in the 1850s.

Martini Action
The generic term now used for the tilting-block (or falling-block) single-barrel, under-lever operated rifle design patented by Friedrich von Martini in 1868. It was adopted by the British Army in 1871 as a service rifle, with Henry rifling and was known as the Martini-Henry. The Martini-action is found in shotguns made by W.W. Greener known as the 'GP'.

Main Spring
The largest, strongest spring in a gun-lock, which powers the hammer or tumbler to achieve detonation. English guns traditionally used leaf-springs but most modern factory-produced over/unders use coil-springs.

Monte Carlo Stock
A stock with a high, straight comb and a drop immediately before the heel., originally designed for trap shooting disciplines and named after Monte Carlo, where many 'live-pigeon' shooting competitions were held.

Non- Rebounding Lock
Early hammer guns had to be placed on half-cock to lift the hammer off the striker and enable the gun to be opened and on 'full-cock' to enable it to be fired. This two-stage cocking system was replaced in the mid 1860s by non-rebounding locks.

Obsolete Calibre
In the UK, certain calibre firearms are considered to be 'obsolete' because ammunition is no longer readily available for them. These are individually listed on the Home Office website and the list is updated for time-to-time. No ammunition is considered

obsolete; only the guns. Such guns can be possessed by members of the public, without the need for any kind of licence.

OFF THE FACE
When a gun has been used for a long period or has been used with excessively powerful cartridges, the space between breech face and barrels may become wider than desirable or the whole jointing between barrels and action may become loose. When this occurs, a gun is termed 'off the face' and must be re-jointed.

OVER & UNDER
A configuration of barrels in a double gun in which the barrels are placed one above the other rather than side-by-side. Of the British makers, Boss (1909) and Woodward (1912) are the best known and most copied designs, though J.M. Browning was the designer whose classic, factory-friendly over/under was the most globally influential.

OUT OF PROOF
The proof marks stamped on a gun by the London or Birmingham proof houses are valid indefinitely, as long as the gun stays in good condition. If a gun comes 'off the face', is badly pitted, or is lapped so that the bore dimensions exceed the original ones by ten thousandths of an inch, it is deemed 'out of proof' and may not legally be offered for sale.

PARADOX
Holland & Holland bought the rights to a patent by Lt. Col. G.V. Fosbury for a double shotgun with rifled chokes firing a conical bullet with acceptable accuracy to 100 yards as well as produce decent patterns as a shotgun. They named this the 'Paradox' and it was in production from the mid-1880s until the mid-1930s. A 12-bore Paradox could put five shots into a 7" square at 100 yards. H&H began Paradox production again in 2006.

PATTERN
Pattern is the term given to the spread of shot delivered by a shotgun. This is typically shown on a pattern plate (of whitewashed iron). An effective pattern should show the pellets are evenly spaced and that there are no large gaps through which a bird may escape or be only wounded. A pattern is usually measured by drawing a 30" circle from the centre point and counting the pellets within the circle.

PATTERN PLATE
A whitewashed metal plate at which guns are tested for pattern and regulated until a desired a pattern is achieved.

PURDEY BOLT
The patent of James Purdey, the 1863 double under-bolt is the default means by which breech-loading double guns are locked shut. The spring-loaded bolt engages with bites in the under-lumps to form a secure lock-up. The Purdey bolt is now almost universally used in tandem with the Scott spindle and top-lever.

PIN-FIRE
A pinfire gun is a breech-loading hammer gun loaded with a cartridge, which has a log metal pin protruding from the side of the metal base. The pin stands proud of the breech end of the barrels, through which a channel is drilled to accommodate it. The hammer falls and its flat surface strikes the pin, detonating the cartridge. Centre-fire took over in 1861.

PIGEON GUN
A gun designed for the sport of live-pigeon shooting. Pigeon guns are heavier than game guns (typically between 7½ and 8lbs) and will usually be proofed to take heavier cartridges (2¾ or 3" chambers designed for 1 ¼oz of shot). They have straight, high combed stocks with pistol grip and long barrels with flat, machined or filed ribs. The barrels will usually be 30" or 32" long and choked full and full.

PIGEON RIB
A flat, straight, raised, rib designed for trap or live-pigeon shooting. The surface is filed or engine turned to reduce glare.

PISTOL GRIP
A stock in which the 'hand' is contoured into a shape reminiscent of a pistol's handle. The pistol grip stock provides a more secure grip and is popular on double rifles and pigeon guns. It is especially suited to guns with single triggers, where the hand does not have to slide back to use the rear trigger for the second shot.

POWDER
The propellant used in shotgun ammunition. Today powders are nitro-celulose based 'smokeless' powders but in the 19th Century black-powder was in general use.

PIGEON MAGNET
Invention of Oxfordshire professional pigeon guide, Philip Beasley, the 'magnet' is a mechanical rotary device for use in decoying pigeon into shot. It consists of an electric motor and two rotating arms, to which dead pigeons are mounted. The movement simulates birds landing to feed and encourages others to investigate.

PELLET
A shotgun cartridge is loaded with individual pellets of a specified size. The most common for game shooting are No.5, No, 6 and No.7.

PRINCE OF WALES GRIP
An elongated form of pistol-grip with a horn or metal grip cap. It is an elegant shape and provides easier movement of the hand to locate the rear trigger than a conventional pistol-grip or semi-pistol grip. It is said to be named after Edward VII, who was Prince of Wales during the reign of Queen Victoria, though supporting evidence for this is hard to find.

PERCUSSION CAP
A cap, generally of copper, which contains a charge of chemical compounds capable of ignition via a sharp blow. The percussion cap is placed on the nipple of a percussion muzzle loader and when struck by the hammer, sends a spark down the hollow nipple to the powder charge, thereby discharging the firearm. Modern cartridges have the percussion cap as an integral part of the base.

PERCUSSION FENCES
Fences shaped with a raised perimeter, as originated in the era of percussion muzzle-loaders. The style was retained for a time by makers of breech-loading hammer guns.

PIT
A deep cavity in the bore of a shotgun caused by corrosion.

PROOF
Proof is the system and the related laws developed to ensure that guns sold to the public are safe to use. All guns must be submitted for testing to legally-specified tolerances before they are stamped and can be sold.

PROOF HOUSE
There are two proof houses in the UK: one in Birmingham and one in London. They are the only bodies legally entitled to apply the approved proof tests and stamp the requisite proof marks on firearms.

PROOF LAW
Proof law in the UK is a civil law governing firearms. It requires that all firearms made and sold to the public are' proofed' as safe by one of the two approved Proof Houses and remain in proof if they are subsequently sold.

PROOF EXEMPT
The Proof houses may, at the discretion of the Proof Master, deem a gun of historic interest to be 'unprovable' due to design or condition and may therefore issue a Certificate of Unprovability. The gun can be sold as a curio and not for use without conforming to proof law.

PUMP ACTION
A gun with a tubular magazine under the barrel which is re-loaded by the rearwards movement of the front hand on the forend.

In the UK a shotgun may hold only two shells in the tube and one in the chamber, otherwise it is considered a 'firearm' and requires a Firearms Certificate for ownership.

PUNT GUN
A large-bore, usually single barrelled, gun, mounted on a low-profiled boat (punt) and used for shooting large flocks of wildfowl on the water. They were mostly used by market hunters but were championed by Sir Ralph Payne-Galwey, who designed a double-barrelled breech-loading punt gun, subsequently made by Holland & Holland. It fired 20oz of lead shot.

REPEATER
A shotgun which fires successive cartridges with repeated pulls on the trigger, with or without some form of manual manipulation. They can mostly be categorized as semi-automatic, pump-action or lever-action.

RE-PROOF
If a shotgun becomes 'out of proof' due to poor condition, enlargement of the bores or modification, it must be returned to the proof house and subjected to proof testing. It will then be stamped according to the proof test and a 'reproof' mark added.

REBOUNDING LOCK
The fired lock of early hammer breech-loaders would come to rest with the hammer resting upon the striker. It required the shooter to pull the hammer to 'half-cock' before attempting to open the breech. Failure to do so made the gun difficult to open because the striker would be stuck in the head of the cartridge. After 1867 most new guns were made with Stanton's rebounding locks, which automatically place the hammer at rest in a position slightly removed from the striker, allowing the gun to be opened without first manipulating the hammer.

ROUND ACTION
The term is properly attributed to John Dixon's trigger-plate action gun of 1880. The gun is well-balanced, strong and reliable. The term is sometimes, erroneously, used to describe sidelocks or boxlocks with a rounded profile to the bar.

RIFLED CHOKE
A section of constriction at the muzzle of a shotgun (usually around 3") that is rifled. It allows a conical bulled, loaded into a shotgun shell to be fired with fair accuracy as well as allowing the gun to be used with shot-shells.

RECESS CHOKE
A system of choking, or replacing choke in shortened barrels by lapping metal away from the bore immediately behind the muzzles to provide a de-facto constriction at the muzzle.

REPLACEMENT BARRELS
New barrels made to replace originals,

which may have been damaged or worn beyond economic repair.

RIB

A side-by-side shotgun has a rib joining the barrels at the top and at the bottom. The bottom rib is insignificant in shooting terms but the top rib may be filed into a variety of profiles depending on the view down the barrels required by the shooter.

RISING BITE

A rising bite is a rising bolt, which engages from below with a hole in a top-rib extension. The best known is the Rigby & Bissell patent but others exist, such as the one used on J.H Walsh's 'Field Gun' patent of 1884.

RIVVEL

A rivvel is a series of tiny bulges in a shotgun barrel wall caused by a partial obstruction or over-loaded cartridge.

RANGE

Range is the effective distance at which a gun can be used. To say a bird or beast is out-of-range means that it is too far away to be shot at with the certainty of a clean kill if the charge meets the target.

RIBLESS GUN

Best known as a feature of certain Alex Martin shotguns. The rib does not extend the full length of the barrels but stops after around 6" and thereafter spacers join the barrels. It was said the loss of rib reduced weight and prevented water entering the rib and rusting the barrels from within.

RIBANDED FENCES

A style of carved fence in the manner of ribbons. Typical of Webley-made Army & Navy guns of the turn of the Century

REGULATE

Regulating a best gun involves hours of time at the pattern plate. The gun will be adjusted until the pattern or bullet hits the desired point of aim at a given distance. A shotgun is also adjusted until the desired percentage of pellets is evenly spread in the 30" pattern at the prescribed distances for each barrel.

RATIONAL STOCK

The 'Rational Stock' was a design attributed to W.W. Greener. It is an alternative style to the Monte Carlo stock and is said to give better and more regular sighting alignment than more traditional stock shapes. It is encountered only rarely.

SHOT

Shot is the term used for small spheres of lead alloy used in shotgun cartridges. Shot is formed in a Shot Tower. Molten lead is poured through a type of sieve to regulate the size of droplet. It then falls through chilled air into water, where it hardens. The most

common shot sizes for shooting game in the UK range from BB for long-range wildfowl shooting in large bore guns to No.8, which may be used for shooting snipe. No. 5, 6 and 7 are the usual sizes for most game shooting.

SHOTGUN

A gun of smooth bore gun with a barrel not less than 24", according to British Law. A wider definition does not restrict the barrel length but refers to a gun designed to fire cartridges containing multiple pellets rather than a single projectile.

SPRING COCKING

In 1880 Frederick Beesley designed a sidelock which is cocked by one arm of the mainspring when it is closed and also operates as a self-opener, again using the mainspring to power the movement; a design associated with Purdey sidelocks ever since.

SEMI AUTOMATIC

A gun which chambers the next round without any action from the shooter. The first successful semi-auto was Browning's A5 of 1905. They are either operated by recoil (like the A5) or by the blow-back of gas from the fired shell (like the Remington 1100).

SPLINTER FOREND

A fore-end of very slim profile, found on a side-by-side. This is the traditional style for a British double game gun. Live-pigeon guns and American shotguns are often found with a wider 'beavertail' forend. British shooters generally wear a leather glove on the forward hand as the grip is as much on the barrels as it is on the wood of the forend.

SEMI-PISTOL GRIP

A shaped hand to a gun stock which is less steep in drop than a full pistol grip and the lower-most part of the grip is rounded rather than squared or finished with a metal or horn grip cap.

SIDE LEVER

An operating lever extending from under the action to the side of the lock-plate. When pressed down the side-lever disengages the bolts and allows the barrels to fall. Sidelevers were widely used on lower quality guns well into the 20th Century but are best known on 'best' guns by Stephen Grant and Boss.

SNAP ACTION

An action which closes and bolts automatically (via a spring) without manual manipulation of the lever. The raising of the barrel is all that is required. Snap actions are faster to operate than the earlier inert actions of the 1860s.

SCREW GRIP

Webley & Brain patented a screw-grip located on the top-lever, which secured

a top-extension of the rib to the action. It is always found in tandem with a Purdey under-bolt and was widely used on Webley boxlocks made for the Trade. 'Screw grip' was earlier used to describe the inert rotary under-lever design of Henry Jones (1858)

SHELL FENCES

A style often encountered on the guns of W.W. Greener in which the fences are carved in the shape of sea shells.

STOCK

The wooden part of a gun located behind the action.

STOCKER

The specialist gun maker trained to make and fit the stock to the action.

STOCK OIL

A mixture of linseed oil, driers and waxes used to impart the waterproof layer which gives a gun stock its shine. Applied by hand over several weeks.

STRIKER

The small, pointed piece of metal which connects with the cartridge cap to instigate ignition when struck by the tumbler. Some boxlocks have the striker as an integral part of the tumbler.

STRIKER HOLE

The small hole, or holes, in the breech face through which the striker emerges when the gun is fired.

SEAR

A sear is a metal limb which is engaged with a notch (bent) in another limb within the lockwork. When the trigger is pulled, the trigger sear is moved backwards, out of the bent in the tumbler (having been held there by spring tension), allowing the tumbler to fall and fire the gun.

SOLID RIB

A rib with no gaps or spaces between the top of the rib and the barrels. The feature is usually associated with over/under guns. It is an alternative to a ventilated rib. The mid rib on some over/unders may also be described as solid if it has no gaps or spaces from breech to muzzle.

SKELP

A type of twist barrel made in the midlands. Skelp was a one-iron twist barrel of undistinguished figure but perfectly serviceable and used in lower-priced guns.

SKELETON BUTT PLATE

A metal plate covering the butt sole of a gun. Rather than being solid, it has been punched or cut so that only a frame and cross-pieces, often of a decorative nature, remain.

STUB IRON

A type of twist barrel made for the stubs cut-off during the production of horse-shoe nails. These stubs were

heated and hammer-welded into a rod, which was in turn hammer-welded around a mandrel to make a barrel tube.

SLEEVING

A method used to restore worn gun barrels. The old barrels are cut just beyond the chambers and new tubes sweated into the old chambers.

SLACUM

A gun maker's term for stock-oil.

SIDE SAFETY

W.W. Greener was noted for placing the safety catch or slide on the side of his boxlocks rather than on the top strap, as was (and is) the convention. His contention was that it removed less wood from a crucial area needing strength and consequently made for a more robust gun.

SIDE PLATE

Some boxlocks have side plates of metal applied to give the gun the external appearance of a sidelock. Many modern trigger-plate over/unders also employ this tactic to improve a gun's appearance.

SIDE PANEL

Some early boxlocks of quality featured a checkered panel behind the action as a decorative feature. They are especially prevalent on Hollis boxlocks and early Westley Richards boxlocks.

SCOTT SPINDLE

The patent of William M. Scott, the Scott spindle was mated with the Purdey double under-bolt to form the classic means by which to open and close a double shotgun. The spindle is operated by a top-lever and connects with the Purdey bolt by means of a vertical hole drilled into the action body, between and just behind the fences.

STRIKE UP

Striking-up is the term used to describe the process of smoothing and profiling barrels before blue or brown rusting is applied

SEXTUPLE GRIP

The first Purdey over/under was a very strong but rather cumbersome design based on the 1912 patent of Edwinson Green. To add strength, six bites were provided to lock the gun shut. It sold poorly and was later reduced to four bites before being discarded in favour of the J. Woodward o/u that Purdey still makes.

TOP STRAP

The top strap is an extension of the action. It extends some way down the hand of the stock and is frequently used to carry the safety button on a hammerless gun.

Top straps on double rifles and many shotguns of the 1860s and 1870s were longer than the current norm.

Some designs incorporated lock-work mounted on the top strap.

TUMBLER

When hammerless guns replaced hammer guns, the term 'tumbler' replaced the term 'hammer'. Essentially a tumbler is an internal hammer.

TUMBLER PIVOT

The tumbler is mounted on a peg, which extends through the lock plate of a sidelock, or through the action wall of a boxlock. This tumbler pivot is a weak spot on boxlocks. On sidelocks it is often protruding and used to carry a cocking indicator.

TRIGGER

The trigger is the conventional means by which the shooter's finger operates the firing of the gun. This is generally by pulling it backwards. Some competition guns have triggers that are pulled but do not drop the tumbler until they are released.

TRIGGER PLATE

The trigger plate is the metal plate, attached to the action by means of strong pins (screws) upon which the trigger mechanism is attached. Some guns (notably the 1880 Dickson 'round action') incorporate the lock work on the trigger plate.

TRIGGER PLATE ACTION

A gun in which the lock work is mounted on the trigger plate rather than on side plates. The best-known are the 1880 Dickson 'round action', the 1879 MacNaughton 'Edinburgh Gun' and the 1873 Phillips patent built by Stephen Grant.

Trigger plate actions are now popular in over/under designs like those of Perrazzi and Beretta and continue to be developed by gunmakers. In 2006

Abbiattico & Salvinelli and Purdey both announced newly designed trigger-plate guns.

TRAP

Originally the trap was the basket in which a live bird (usually a pigeon) was kept and released from during live-bird shooting competitions. Since the introduction of inanimate bird (clay pigeon) shooting the term has been used for the machine used to launch the clay targets.

TRAP GUN

A gun designed especially for shooting competitions involving either live-birds released from traps in front of the shooter, or for clay targets simulating this type of going-away shot.

Trap guns typically have a point of impact a little above the point of aim and long barrels. Other trap gun features include pistol-grip stocks, raised, matted ribs and heavier weight than a game gun.

TRADE GUN

The term used to describe guns made by large factories and supplied to smaller concerns for retail or for finishing. A Webley-made boxlock retailed by a provincial gun shop is typically described as a 'Webley trade gun'.

TOP LEVER

The preferred means by which breech-loaders are opened. The early patents like the Westley Richards 1863 patent used a locking bolt which engaged with the rib extension, others operated single bites in the lumps.

However the most successful top-lever operated system is the combination of top lever, Scott spindle and Purdey double under-bolt.

TWELVE TWENTY

The patent of Birmingham gun maker William Baker, the 1906 back-action operates as an assisted opener and can be built lightweight but strong due to the solid bar. This Baker action was used by Churchill, Grant and William Powell but is associated most famously with Charles Lancaster.

TOE

The lower-most extremity of a shotgun butt sole. It fits close to the armpit of the shooter as a shot is taken.

TREBLE GRIP

A system of locking a gun by means of Purdey double under-bolts and a Greener cross bolt through a top-rib extension.

UNDER-LUMP

The lumps on a breech-loader are commonly found to be projections emerging from the barrel flats. They can be quite varied in form and number, usually from one to four. The forward lump forms a pivot with the hinge pin and bites in the rear of the lumps are often used as a means to bolt the barrels closed.

VICKERS STEEL

When forged steel emerged for use in barrel tube production as an alternative to Damascus, several industrial names became involved and the barrels of these early (1890s) guns are often found to bear the name of the steel maker. Vickers tubes were among those used and this explains the term 'Vickers' stamped on the barrels of such guns.

WHITWORTH STEEL

Sir Joseph Whitworth was a giant of Victorian industrial engineering. In the 1890s he developed a system for

producing solid steel forgings suitable for use in barrel tube production. This involved hydraulic squeezing of the liquid steel as it cooled, thus removing the air-contamination from the centre of the forging. Such forgings could then be drilled to form barrel tubes. Many gun makers offered the (very expensive) Whitworth tubes and they will generally be stamped 'Made from Sir Joseph Whitworth's fluid-pressed steel' as well as bearing his trademark wheat-sheaf stamp.

WOOD-BAR ACTION (or BAR-IN-WOOD)

A style of breech-loader in which the stock partially covers the metal bar of the action. This gives a profile similar to a muzzle-loader: the wood extending seemingly un-interrupted from butt sole to forend finial. A style commonly found in hammer guns up to 1880s.

WALNUT

The preferred wood for shotgun stocks. Walnut is strong, dense and easy to cut. It is also stable and hard-wearing. The preferred location for sourcing gun stock walnut was once the south of France. However, stocks are almost totally exhausted there and most British gun makers now use walnut from Turkey.

XXV

Robert Churchill used the 'XXV' logo to distinguish his 25" barrelled shotguns from around WWI. He developed a shooting style for the guns and a distinctive raised and narrowing rib to match. The XXV was built in a range of qualities, both boxlock and sidelock.

ZENITH

The Zenith was an over/under model developed by Churchill but it was never commercially successful.

Wherever you shoot and whatever you shoot with – appreciate the social side of shooting: the friendships, the memories and the company of fellow sportsmen.

Cheers!